CLIMATE ZERO HOUR

A plea for sanity in the Energy debate.

Written
by
Mathijs Beckers

"I hope I have helped to raise the profile of science and to show that physics is not a mystery but can be understood by ordinary people."—Stephen Hawking

Acknowledgements

Cover Art:

Robyn Gough

Proofreading & Editing:

Meredith Angwin	Alan Medsker
Dr. Alexander Cannara	Erik Schmitz
Canon Bryan	Todd de Ryck

Dr. Ripudaman Malhotra

Preamble

With this book I'm not contesting the validity of climate & ocean change science, far from it. The need to act becomes more urgent with each passing day. It is said that we have all the tools to help mankind rise above the dependence on fossil fuels. However, the manner in which we could reach this elevated state is in dispute. Some think that we could reach a low-carbon future by embracing the principles of the renewable technologies, while excluding other possibilities like nuclear energy. On the other hand, we note that there exists a strong contingent of academics and intellectuals who embrace the combination of as many feasible solutions as possible and are unwilling to commit fully to just one set of technologies while ruling out another.

These opposing sides are intertwined in a tug-of-war. One side has made fear a guiding principle. Fear of losing face and popularity and credibility, and by extension, funding. Whereas the other side, if we may call it that, has science and reason as their guideline. The key focus is to determine what is truly feasible when witnessed by our descendants, and how we will be judged by them only a couple of decades from now.

I am going to show you how the discussion is unfolding. And what tactics are used to influence the public. Also, I would like to provide some insight in the "Inclusive versus Exclusive" dichotomy that has been placed before us.

In conclusion, I believe that the "Exclusionists" are counting their chickens before they have hatched. So far, I have not been able to find an argument compelling enough to exclude nuclear and endorse a future powered exclusively by renewables.

Units

Most units in this book are in Metric, Celsius, and Short-Scale; Kilo = 10^3; Mega = 10^6; Giga = 10^9; Tera = 10^{12}; unless stated otherwise.

Prologue

The problem is more urgent than we think

First off, even though I really would want to spend my life studying the issues of our age, I am neither an expert, nor a scientist. This book is the culmination of more than seven years of (nonacademic) research and modelling. I am going to present you with my viewpoints on the ongoing energy debate. It is imperative that you don't accept anything I write, just because it fits your viewpoints. I'd rather have you check and recheck my claims as well as other people's claims.

Mankind has been addicted to fire for over 400,000 years—and it needs an intervention.

Let's start with considering some *"dry"* scientific models and try to understand what they mean; why we must take them seriously.

Several possible temperature scenarios have been established by the Intergovernmental Panel on climate change, or IPCC. We use these models to determine how much the average temperature on Earth could rise when we reach a certain concentration of atmospheric carbon dioxide (CO_2). We reference these values to those at the beginning of the Industrial Age. These scenarios are called Representative Concentration Pathways or RCPs[1]. There are four models which represent the possible solar radiative forcing by 2050, ranging from an additional 2.6 Watts per square meter to 8.5 Watts per square meter relative to pre-industrial level. The radiative forcing in these models is positive, which means that the Earth and its atmosphere capture more energy than is irradiated back into space, and this means that the Earth, air and oceans are warming up[2].

If we continue with business as usual (BAU in scientific publications), we end up in the RCP8.5 scenario. If this happens, it is likely that the Earth will have warmed by 4 degrees Celsius relative to the average temperature at the beginning of the Industrial Age. Even worse, in a recent article in the Nature International Journal of Science[3], by postdoc Patrick T. Brown and Professor Ken Caldiera, it is argued that it is more likely that we will end up in the high end of temperature change predicted by the IPCC's RCP scenarios. Which means that we're faced with higher temperatures, and that the low ends of the IPCC's RCP temperature predictions are less likely to occur.

Additionally, the RCP scenarios attempt to predict how much the sea will rise relative to pre-industrial levels. Major contributors to the change of the sea-level rise are melting glaciers and land-based ice and snow caps. Also, an increase in overall temperature of seas and oceans causes water to expand[4] and its chemistry to change. In the model for business as usual (RCP8.5) we note a possible sea level rise of 0.6 to 0.8 meters. This would be enough to flood significant parts of New York, New Orleans, Calcutta, Miami, the entire southern river delta of Vietnam, great portions of the Chinese Pearl Delta, several Pacific islands, and many of the major coastal floodplains across the planet[5]. This could affect tens of millions, if not hundreds of millions of people. In fact, if we consider this more carefully, we can assert with a high degree of certainty that whatever happens at the coasts, will inadvertently affect the entire population on Earth. The economic damages resulting from the loss and damage of real-estate and goods from storms will be incalculable; the social costs, potentially catastrophic. US officials have determined climate/ocean change to be a national security threat.

Anthropogenic climate & ocean change is the most potent catalyst for causing societal instability and damage to the ecosystem. This is a broad statement, but it is true. Man-made (anthropogenic) climate and ocean change affects all life on earth. Ranging from the top of the food pyramid, all the way down to its fundament on land and in the seas. It is a sad fact that intermediate collapses are already happening. We see collapses in our own civilization due to crop famine and declined water availability in unstable areas; we see collapses in nature, where species are going extinct[6]. Do note that extinction is permanent, and we're to blame. The current rate of man-induced extinction is only rivaled by prior mass-extinction events[7] that wiped out more than half of all the life at those epochs on earth, five times before our own.

Is it possible to quantify the damages? It's more than just the warming of the earth. We've destroyed habitats, we've hunted animals to the brink, and we've polluted great areas of our planet, rendering areas uninhabitable. Not to mention that great areas of the oceans are now being acidified, warmed, even deprived of oxygen, causing massive hypoxic dead zones, where life simply cannot exist[8,9].

If we consider carbon emissions of the past decade, we see a clear increasing trend, without any indication of slowing down[10]. This is worrying, as this confirms that mankind is going on with business as usual. However, this is not entirely accurate; the OECD has managed to drop some of its emissions, but only to a very small degree[11]. The developing countries are catching up, in terms of energy availability per capita, at a great pace, and none of this is possible without generating enough energy and resulting in unavoidable emissions that come with the combustion of fuels[12].

The problem is the rate of growth. We have been emitting carbon at an increasing rate, and this means that it is becoming increasingly more difficult to stop the climate and oceans from changing so much that we will leave a much less hospitable planet to those who come after us. Additionally, we must account for inertia of the Earth's atmospheric and hydrological systems[13]. They can capture a lot of greenhouse gases and energy before we notice much change. We've started using fossil fuels at an accelerated rate early in this century and have been dragging our feet on the job of overhauling our energy infrastructure and combatting anthropogenic climate change for over 20 years, since the Kyoto Treaty[14]. One bright, clean-power example, since the OPEC embargo, has been France (2015, "France emits around 40 grams of CO_2 per KWh... most other industrialized nations emit between 400 and 500 grams of CO_2 per KWh[15]).

To give you a sense of urgency, we will now consider one of the possible effects from our carbon emissions. We will skip over the possibilities of oceanic overturning circulation shutoff, desertification, droughts, crop failures, famine, increased pests, vector-borne diseases, dying forests, disappearing glaciers, an ice-free arctic, methane release from clathrates, methane release from rotting vegetation thanks to dissipating permafrost, irregular and extreme precipitation patterns, more powerful storms, sea level rise, and dive straight into the ocean—to look at it from a biologist's and chemist's perspective. Note that I published the following section before in the *"Non-Solutions Project."* For me, it's the best way to explain this problem.

At the ocean's surface gases are exchanged between the atmosphere and the water[16]. Some come out of solution, and some go into solution. Among these, is CO_2. Somewhere between a third and half of the carbon dioxide we created has been absorbed by the oceans, where it reacts with water, creating carbonic acid (H_2CO_3) which makes seawater more acidic (less alkaline or basic).

Two reactions determine how many protons (H^+) are released per molecule of carbon dioxide as it reacts with water to create carbonic acid.

$$H_2CO_3 \rightleftharpoons HCO_3^- + H^+ \text{ and } HCO_3^- \rightleftharpoons CO_3^{2-} + H^+$$

Here's where it gets interesting. The following describes the dominant CO_2-sequestration system on our planet. Calcifying sea life takes carbonate ions from seawater to build shells and skeletons, and when dying, takes that carbonate to seafloor sediments and eventually turns into to limestone. These animals permanently sequester about 1 billion tons of CO_2 per year[17]—nothing matches their work.

We have evidence to suggest that this vast amount of excess carbon dioxide has lowered the pH by one tenth of a percent (about 30% more acidic—note that the pH scale is logarithmic)[18]. Another concern is the increased availability of protons to form bonds with the CO_3^{2-} (carbonate) ions, and this is a problem for shell-forming organisms as it means that there is an imbalance between carbonate ions and Ca^{2+} and Mg^{2+} ions which the shell forming organisms need to form calcium & magnesium carbonate. These carbonates are the main building blocks for shells and bones, and therefore essential for life forms such plankton, oysters, clams, krill, whales etc... And warming seawater makes it worse.

Would it be enough to remove carbon dioxide? According to a paper called *"Long-term response of oceans to CO_2 removal from the atmosphere"*[19], it would not be enough, especially if we keep going on with business as usual. In fact, if we keep introducing excess (man-made) carbon dioxide to the atmosphere it will damage marine life, regardless of our efforts to capture and sequester carbon dioxide. And that will probably happen before IPCC temperature limits are reached.

Marine life, including corals have already been damaged[20]. The bleaching of corals occurs when waters become too warm, or when they are being exposed to more sunlight and warming. Ocean acidification keeps coral reefs from growing, as less (or no more) calcium/magnesium carbonate is available for reef creation—which is a byproduct of a biochemical process of tiny single-celled symbiotic life forms called Zooxanthellae[21].

This means that we must focus on the well-being of marine life, particularly on coral reefs and plankton, to keep the marine life pyramid from collapsing. To accomplish this, we must greatly curtail carbon emissions and start capturing and permanently sequestering carbon dioxide—if at all possible...

For more in-depth information, please see marine chemist Andrew Dickson's informative, YouTube presentation called: *"Acidic Oceans: Why Should We Care? —Perspectives on Ocean Science"*[22] and Peter Brewer's *"A short history of ocean A short history of ocean acidification science in the 20th century: a chemist's view"*[23].

How big is this problem?

According to the NOAA the atmospheric concentration of CO_2, per December 2017, is 406.82 ppm. In 2016, in the same month, the concentration was 404.42 ppm.

Consider these facts and figures from the Fifth Assessment Report of the IPCC[24].

"Annual CO2 emissions from fossil fuel combustion and cement production was 8.3 [7.6 to 9.0] GtC yr–1 averaged over 2002–2011 (high confidence) and were 9.5 [8.7 to 10.3] GtC yr–1 in 2011, 54% above the 1990 level. Annual net CO2 emissions from anthropogenic land use change were 0.9 [0.1 to 1.7] GtC yr–1 on average during 2002 to 2011 (medium confidence)."

"From 1750 to 2011, CO2 emissions from fossil fuel combustion and cement production have released 375 [345 to 405] GtC to the atmosphere, while deforestation and other land use change are estimated to have released 180 [100 to 260] GtC. This results in cumulative anthropogenic emissions of 555 [470 to 640] GtC."

"Of these cumulative anthropogenic CO2 emissions, 240 [230 to 250] GtC have accumulated in the atmosphere, 155 [125 to 185] GtC have been taken up by the ocean and 160 [70 to 250] GtC have accumulated in natural terrestrial ecosystems (i.e., the cumulative residual land sink)."

"Ocean acidification is quantified by decreases in pH. The pH of ocean surface water has decreased by 0.1since the beginning of the industrial era (high confidence), corresponding to a 26% increase in hydrogen ion concentration".

Note: the term GtC means metric Gigaton Carbon, which is not equal to $GtCO_2$. One GtC equals 3.667 $GtCO_2$.

Roughly 1/3rd of our unnatural CO_2 emissions have been dissolved in seas, but even that amount has been sufficient to lower their pH more rapidly than at any time since the great Permian Extinction. Even if we stop all CO_2 emissions today, it would not restore ocean chemistry to pre-industrial levels.

The IPCC estimates that, given their best climate change mitigation model, cumulative CO_2 emissions from 2012 to 2100 will be at least 510 Gigatons of CO_2, and this does not include the ~2035 Gigatons of CO_2 we have emitted since 1750.

According to the IPCC ocean acidification will look as follows according to these RCP Models:

"Earth System Models project a global increase in ocean acidification for all RCP scenarios. The corresponding decrease in surface ocean pH by the end of 21st century is in the range of 0.06 to 0.07 for RCP2.6, 0.14 to 0.15 for RCP4.5, 0.20 to 0.21 for RCP6.0, and 0.30 to 0.32 for RCP8.5 (see Figures SPM.7 and SPM.8)."

"pH is a measure of acidity using a logarithmic scale: a pH decrease of 1 unit corresponds to a 10-fold increase in hydrogen ion concentration, or acidity."

According to the Smithsonian Institute: *"If the amount of carbon dioxide in the atmosphere stabilizes, eventually buffering (or neutralizing) will occur and pH will return to normal."* However, they also state: *"But this time, pH is dropping too quickly. Buffering will take thousands of years, which is way too long a period of time for the ocean organisms affected now and in the near future."*

It should be clear by now that we must ring the alarm bells on ocean acidification and start thinking about reducing the amount of CO_2 in our atmosphere and oceans to a sustainable level, as well as for the rest of the biosphere. It is said that a CO_2 concentration of 350 ppm is sustainable. Please see *"Target atmospheric CO_2: Where should humanity aim?"*[25] to learn why we must try to get below 350 ppm while simultaneously trying to protect ocean chemistry.

We're going to end this chapter with a stark observation:

"Make America America again"[26] by James Hansen, as excerpted from his letter called "How does it feel?" which was written on October 4th, 2017:

"Our two major political parties are competing to see who can do more damage. In Sophie's Planet I will argue the need for a new, centrist, party, with the objective to "make America America again." That sure won't happen with either of our present two elitist political parties, both of which have gone off the rails. Representatives of the party dominated by deniers, are honest crooks. They don't hide the fact that they are on the take from the fossil fuel industry. Science be damned; it's all about money.

History may find the other party to be more destructive. They fool the public and themselves. They are the Neville Chamberlain party. They pretended that the Kyoto Protocol would do something. Now they pretend that the Paris Agreement does something. Kyoto and Paris are analogous to what Churchill described as "half-measures," and "soothing and baffling expedients." As a result, young people will be "entering a period of consequences".

Have you heard the hogwash about the world turning the corner, moving to clean energies, phasing out fossil fuels? You heard it in 2015 with all the politicians clapping each other on the back in Paris. You can read it daily in "Big Green" propaganda machines, such as EcoWatch, just to pick on one of them."

Given our apathetic response to this looming calamity, I wonder…Are we too late? How much damage have we already caused? And how much is still in the pipeline? Our planet and its different systems have a lot of inertia, and we're feeding it with an increased amount of energy. I'm certainly very anxious to see where we are headed. And I wonder whether we can set aside our differences and create some consensus on how to move forward.

I'll leave you with the title of an article, written by Eduardo Porter and published in the New York Times on January 23rd, 2018:

"Fighting Climate Change? We're not even landing a punch."

Part one: Climate Change Split

Let me preface this chapter with the fact that I've been going back and forth between pulling no punches and trying to build bridges while writing this part. This conflict is reflected in my writing style. On one hand, I keep hoping that some of these individuals will finally come to their senses. On the other hand, I want you to be outraged about the continuous 100% renewable / anti-nuclear demagoguery that is going on. Not to mention the odious juxtaposition these people invoke. They make it seem as if nuclear and renewables are opposing technologies, with those in support of these technologies being in equal opposition. These assertions are false. Here's the correct perspective: nuclear and renewables aren't mutually exclusive. In fact, I think that they are complementary to each other, and so do others. But we will learn more about that later in this book.

This chapter will make it seem as if I am in opposition to renewable because I will be addressing the talking points of a couple of individuals from a critical viewpoint. Do note that I am not against renewables at all. I am just opposed to the incorrect idea that using renewables exclusively will ensure a steep enough decarbonization curve while simultaneously providing a solid basis for a modern future in prosperity, safety, and stability for all.

Consensus (it's not what you think)

Consensus: A generally accepted opinion or decision among a group of people.

As far as I can tell, there's a broad consensus between all the people who will be subjected to scrutiny in this book. All of them are on board with the scientific evidence that indicates that we are changing the climate, and that the rate of change is too fast and may lead to irreparable damage to the biosphere, of which we are a part, and without which we cannot survive.

However, a consensus on how to move forward has not been reached. Strange things tend to happen when we mix in different ideologies and conflicting viewpoints.

All are thinking about how we can solve this problem. How to put the brakes on a changing climate and its effects on the biosphere. Although, I question whether some have the same sense of urgency. We don't have the luxury of picking and choosing. The renewables-only movement is being spearheaded by academics like Mark Z. Jacobson, who is a Professor of Civil Engineering at Stanford University in California. He thinks we can solve this problem by deploying solar and wind energy, almost exclusively (with a smidgen of hydro, geothermal, solar thermal and other water-based energy technologies). Following in Jacobson's wake we find presidential candidates, other politicians, scientists, science-guys, and Hollywood stars.

The trouble is that Jacobson's thesis has been peer-reviewed. Which means that it can claim to be valid predictive model for feasibility. It answers the question whether we could decarbonize all our energy production and consumption (up to a limited amount) by electrification and by means of wind and solar power (and a very small margin of other renewable technologies). Recent studies have shown that there are severe shortcomings

12

in Jacobson's model, thus putting its feasibility in doubt. But here's the thing: this model doesn't have to be the only model used in our quest for solutions to anthropogenic climate change.

Let's get back to the subject of consensus. We're talking about generally accepted opinions within groups. It is the consensus within great parts of the environmental community that Jacobson's thesis is the best answer to the renewable future they want to achieve. The movement feels strengthened by reports of minor successes in countries like Germany and Denmark and Scotland where, on a very bright or windy day, after investments of billions in local currency, renewable energy accounts for 70 or 80% of their electricity generation. But only a handful of days per year. So, the natural response one encounters whenever you try to explain that renewables alone aren't enough is that you are a denier or a naysayer or a paid shill. These are simple and inane dismissals from people who are cognitively closed to reason.

Headlines they have… Scientific predictive capabilities, they have not.

Parts of the environmental movement are so vehemently opposed to nuclear energy, that they will highlight as many problems as possible to oppose the energy source altogether. Worse, some members of the movement are on record, saying they don't want these problems solved. That's a serious issue. Because this means that, as long as the anti-nuclear faction is pandered to, the environmental consensus won't change to accept nuclear energy as part of the solution. No matter how good our solutions to their perceived problems are, the overall movement won't accept them. If there are significant portions of this movement that would accept nuclear as part of the solution, that could lead to a schism. This would effectively break the environmental movement as it exists today. At least, that's what I suspect. But before anything like this happens, we will have wasted precious time. Just to be clear, I hope that the environmental movement will split, and that a large contingent will accept nuclear as part of the solution. Reason and science, after all, should be our guidelines. I just don't want the world to waste time until the environmental movement becomes truly environmentalist.

Back when populism wasn't yet such a big problem, and long-term energy solutions were still favorable, politicians didn't mind greenlighting long-lasting nuclear projects. Consider for instance France, which decided to decarbonise more than 70% of their electricity by replacing coal and oil with nuclear energy stations. Today, given a much more volatile political climate, in which politicians are always on the hot seat, they tend stay away from subjects that may cost them an election. When we're talking about nuclear energy, this is not so much a problem on the right spectrum, nor in the center, but it is on the left. Especially amongst vocal environmentalists. Although, Finland is one of the few countries were nuclear is an accepted solution, even amongst environmentalists.

At this moment it is unimaginable that a prominent politician like Bernie Sanders would say anything positive about nuclear energy. In fact, in his video conversation with Bill Nye (early in 2017) he said: *"Not to mention you have to get rid of that [nuclear] waste safely."* Which is one of the most often heard objections. Also, I don't think it strange that he doesn't welcome nuclear energy when we consider the great movements that back him up. These, generally, consist of left-wing activists who have adopted climate activism as a social issue, in which it is directly implied that renewables are the only valid option, and this sentiment is strengthened by case studies that show that it can be done under certain circumstances, such as the Solutions Project by Jacobson.

First, Climate Change is not a politically colored issue, it's a scientific matter. It is certain that we are influencing it, and that the effects that we may experience from our influence can be severe. Just because people deny it, doesn't meant it doesn't exist. Sticking your head in the sand doesn't make it go away. It does not logically follow that mitigating the damage from climate change can only be done with deploying a single set of technologies exclusively. But that's what the environmental movement seems to be going for. Even if there are case studies that show that it might be possible using renewables exclusively, doesn't mean that these are our best options. In fact, the amount of permutations that will lead to decarbonization are innumerable. Especially when there's a plethora of viable options at our disposal. But these are the conclusions drawn by hordes of people who have been inundated by the modern environmental herd-mentality.

Luckily, there are now signs of people crossing ideological borders and into the realm of science regardless of previously held convictions. Data trumps tribalism. Consider for instance Bob Inglis, a conservative Christian politician from the US. No one would blink an eye if I told them that he disbelieved climate change and didn't want to act on it. While in fact, he did. However, during a trip to Antarctica, he got a unique insight into how the data (particularly carbon dioxide levels in ice cores) was collected, and he did what any rational human being would do. He re-examined his reasoning and adjusted it to accommodate this new-found confidence in the facts. As a hardline Republican who was elected six times in a district that was equally hardline Republican, his acceptance of the facts came back with a vengeance. As soon as he enunciated his fears about climate change, and his resolve to act upon it, his tribe decided to ostracize him. By accepting the facts, he dug his own –Political– grave. Inglis was no longer a consistent member of the climate-change denying tribe of Republicans.

Also, we have a notable climate scientist, James Hansen. Known as the grandfather of climate change, he is at the forefront of climate activism. I consider him a leading individual, simply because of his expertise in Climate Modelling. Hansen's arguments are based on solid fact, and he is actively seeking solutions, and trying to educate the public on what is going on, and what needs to be done. On occasion I try to get some feedback from him directly, and he is always kind enough to respond.

Bill McKibben founded 350.org, a climate NGO. Its mission statement: *"350 uses online campaigns, grassroots organizing, and mass public actions to oppose new coal, oil and gas projects, take money out of the companies that are heating up the planet, and build 100% clean energy solutions that work for all. 350's network extends to 188 countries."*

McKibben chose this name because of this paper: *"Target Atmospheric CO2: Where should humanity aim?"*[25] authored by James Hansen et. al. And to cap it all off, Hansen was one of the leading scientists who actively supported the 350.org cause. If we consider the 350.org website, we don't find any clue of them being in opposition to nuclear energy. However, they are fervent advocates of transitions to 100% renewable energy scenarios.

Hansen is Director at the Climate Science, Awareness and Solutions (CSAS) group at the Earth Institute at Columbia University. Considering the organizations supported by CSAS, we find 350.org, Citizens Climate Lobby and Our Children's Trust. Three wonderful movements that are clearly involved in mobilizing the government and the people to do something about climate change.

Fear of splitting the movement.

Strangely enough, Hansen has drawn the ire from 350.org board members Bill McKibben (also founder) and Naomi Klein because of his plea to include nuclear energy into future carbon-free energy strategies, thereby effectively pitting the consensus against him. Prominent people now say something like *"I love James Hansen, but he's wrong on nuclear energy."*

Bill McKibben is on record saying: *"If I came out in favor of nuclear, it would split this movement in half."*[28]

Which begs the question: What is more important? Intellectual honesty and fighting for a real chance to save the biosphere? Or, biting your tongue and having a big but ineffective movement? Precisely if someone like Bill McKibben would proclaim that nuclear must be part of the solution, it would change the consensus, this much is certain. Yes, there would be thousands of people who would have great difficulty adjusting to this new paradigm, but many people would change their minds. That not all the people would change their views on nuclear energy is unimportant because we've already pushed the needle into the red on climate change. Greenhouse gas emissions continue to rise; Our population is growing; We're still burning masses of fossil fuels and have not managed to capture and sequester any meaningful volume of carbon from the atmosphere or the ocean.

Bill McKibben is on record saying: "Winning slowly, is the same as losing"

Perhaps McKibben should reexamine his reasoning. If creating new bipartisan support for nuclear would create a significant shift from coal to low-carbon energy sources, then that would be worth splitting an environmental movement in half. Why? Because the result matters. We are not winning slowly. We're still losing significantly. Please Bill, stop this! We will embrace you with open arms and support you vehemently; we will provide you with a chance at becoming a true ally for the environment. I'd bet that James Hansen would be relieved and wants to collaborate with you on spreading climate awareness and pushing forward to create viable solutions that will be carried widely and will lead to actual implementation on a meaningful scale.

Against Centralized Power Generation and Big Industry

Let's consider Naomi Klein's disagreement with James Hansen, which was recorded at the Hammer Museum, on September 30th, 2015. The host of that evening asked Klein whether she agrees with Hansen's view that nuclear should be an essential component to averting climate catastrophe? Klein had a lot to say.

Klein says: "Now, I love James Hansen and I hate to disagree with James Hansen. We all owe him such a tremendous debt of gratitude. He is the real The Godfather of this movement in so many ways. But, yes, I disagree with him on nuclear, and you know I understand why people looking at the current power configurations, as they are, believe that we need these centralized solutions that are less threatening to our elites."

Here, we uncover the first two pieces of narrative that has been adopted into consensus. It goes like this: *"We don't like centralized power generation, we want to have a distributed power system."* And the corollary *"We don't like centralized power generation, because it only serves the elite."* When you engage with environmentalists on these issues, these are the arguments they will present. Naomi Klein is one of the flagbearers of that movement. Do note that she's not an economist, or a scientist, but a social activist.

Klein continues: ". I mean I think the reason why nuclear is seen as a more practical option, is not because we can't get to a hundred percent renewable energy, we can. I mean there's amazing research out of

Stanford by Mark Jacobson that says we can get a hundred percent renewable energy with the existing technologies by 2030."

It is not true that Jacobson thinks we can get to 100% by 2030, it's 2050, a discrepancy of 20 years with Klein's claim. It's probable that she made a mistake. Also, like I said before: *"It is the consensus within great parts of the environmental community that Jacobson's thesis is the best answer to the renewable future they want to achieve."* Bill McKibben, Bernie Sanders, Bill Nye and a couple of other prominent people often use this argument. It sounds like this: *"We can decarbonize, if only we had the will to do it, we only need wind and solar, the Solutions Project says so."*

Klein continues "Okay the problem is that renewable energy is quite challenging to existing power structures, because it's inherently decentralized. The thing about both fossil fuels and nuclear or any extractive industry, is that it is intimately tied to our unequal economic system, because you have resources that are buried beneath the earth, that take a lot of money to get out; a lot of money to refine; a lot of money to transport; and then, in the case of nuclear, a lot of money to deal with the waste."

A couple of important elements are mentioned here that have become part of the consensus.

First, existing power structures are *bad*, because they are centralized. There's a tendency to conflate existing power structures with carbon emissions. But that's not necessarily so. Consider for instance Ontario's power system, which almost entirely decarbonized thanks to hydro and nuclear. Or consider France's power system, to which the same applies. Or Norway's and Sweden's power systems, again, the same applies. These power systems offer a great amount of decarbonized electricity, at reasonable cost. The people of these countries are not oppressed (far from it, they live meaningful and free lives), and these power systems ensure economic stability, and provide tens of thousands of meaningful and good paying jobs.

Second: *"if it requires an extractive industry, it will cost a lot of money."* Third: *"nuclear waste costs a lot of money."* These two elements will be addressed later in this book. Suffice to say, these are not valid reasons to oppose nuclear, as they can directly be aimed back at renewables. We need to compare the volumes concerned to get a clearer picture. Klein's weak assertions in this regard will be disputed with facts.

Klein continues: "Okay so that is a system that's going to lend itself to monopoly power to a few small players, to a corporatist structure between corporations of government."

Corporations, government, monopoly. Three elements accepted by the consensus as bad. But are they? Corporations (social or capitalist) are required to create any energy infrastructure. And how is it possible that government in this regard is a negative thing? The consensus is that it is the wrong kind of government that is bad. Klein does not mean a libertarian viewpoint in which all government overreach is bad. On the contrary, they want to get influence in the government, to enact the changes, they deem required. But this begs the question: are your proposed changes effective enough?

Klein continues: "So, for people, who are defending that very profitable status quo, it's a lot easier to switch from fossil fuels to nuclear, than it is to switch to renewable energy, which is a system that is that is trading in free stuff. Free Wind, free Sun, free waves, that's everywhere. So, you know it's not that money can't be made, but you're not going to make the kind of stupid money that you make from fossil fuels off renewable energy, and that's why it's so threatening."

I have not seen many people make the argument that it would be easier to switch from fossil fuels to nuclear. In fact, I question the validity of this assertion. Other than fitting in a grid, like all other energy sources do, and having the generation side in common with conventional fossil fueled power stations, there's no easy cross-over. The infrastructure required to run civilian nuclear reactors is completely different. Additionally, consider the portfolios of practically all energy companies in the world. They invest in renewables as well. So basically, the argument that it would be easier to switch from fossil fuels to nuclear is moot. Most energy corporations already know how to do this, and there are many precedents to prove this assertion.

19

Also, we get the *"free"* argument. Yet another point incorporated into the climate consensus. *"All the energy we need is free."* That might be true. However, building the infrastructure and the means to capture that energy and make it usable doesn't come for free. *"because you have resources that are buried beneath the earth, that take a lot of money to get out; a lot of money to refine; a lot of money to transport; and then, in the case of solar and wind, a lot of money to deal with the waste."* To use Naomi Klein's own words. If you think that the volume of materials required to build a 100% renewable energy landscape is smaller than the one including nuclear plus the fuel burden that comes with it, I have a bridge to sell you…

Klein continues: "You know, if you have a feed-in tariff and people are able to put solar panels on their roofs, and feed into the grid, those people are energy producers. They're not just energy consumers. Each and every one of them is a competitor for a traditional utility. So, obviously they're going to fight that model."

We add *"everyone can be his own power company,"* to the consensus.

Ascribing motives to all

Klein continues: "Now I understand why somebody who spent their life working for NASA, as opposed to me having spent my life working in social movements, believes that were screwed enough that we have to go for these centralized solutions that are less threatening to the status quo."

Patronizing, that's the best word to describe what Klein is doing here. First, as far as I know, Hansen does not want us to build nuclear power reactors exclusively. In fact, he is on record, telling people that we need all low-carbon energy sources to compete on their own merits. Also, Klein is creating a non-sequitur. It does not logically follow that if you work for a big institution like NASA, that you automatically prefer corporate solutions to a world-spanning problem. And that's not necessarily true. I find it highly unlikely that you could gauge someone's preference for centralized or decentralized solutions based on their working past. I think you should always ask the question personally instead of assuming the answer to the question. Also, let's not forget that it is unavoidable that we're going to depend on corporations and governments to bail us out of this mess. To suppose that it can be done without them, is foolish. It almost sounds like a Libertarian-fairytale... and I don't suppose that Klein is a Libertarian.

Klein continues: "I'm throwing my lot in with social movements and I even understand why a lot of scientists believe the geoengineering is the only way. If we look at where we're at right now and we believe we can't change the political configurations, then it makes sense."

Enter one of the main motivations: to keep the movement as big as possible, and therefore keep the consensus. The idea that this fight can be won in the realm of politics is inherently dangerous. Because now they have pitted themselves against one of the very few solutions that might get bi-partisan support. This is betting the farm on fickle politics. It is vain to suppose that you can select the President and the house. This sentiment also bars them from finding common ground. Consider for instance the plan of the Secretary of Energy (Rick Perry) to create new subsidies for coal and nuclear. All the movement can do in this regard, is fight the very notion. Whereas if they had been less rigid on nuclear energy, they could have reached across the aisle and perhaps reach a compromise. That possibility is now gone. Luckily, on January 9th, 2018, the regulators decided not to go through with this plan. And that's a good thing, because nuclear energy should be the end of coal, not its ally.

Let's also keep in mind that today's political sentiment will change. We might not like most of the GOP representatives, or those with executive functions. That doesn't mean that this cannot change in the future. Keep the avenues open. Don't burn all your bridges.

Social movements to the rescue?

Klein continues: "Why would you be looking at nuclear and geoengineering? But precisely because these are so high risk, that's all the more reason why it's up to the rest of us to build the kind of broad based social movements that can change those structures of power."

Add another element to the consensus: *"We don't like risky things, especially no nuclear risks."* However, unsound the reasoning behind that argument may be.

Klein continues: "So, I guess what I'm saying is I understand, but from my perspective the problem is not just fossil fuels. The problem is an extractive mindset that creates sacrifice zones. I think that at the heart of this, this is not just about fossil fuels. This is about the logic that made us believe we could build our economies on a toxic system that has always been based on sacrificial places, and sacrificial people. Coal was never clean and nuclear demands the same of us once again."

Here it becomes a little bit weird for my taste. I understand that mining in general depends on permanently transforming large swathes of land, kicking people off, mass-denudation, and subsequent legacy damages. This is an accepted element of the environmentalist consensus. It is important to decouple fossil fuels from nuclear fuels. Those two are completely different in orders of magnitude. We're talking about factors of millions in difference. Also, we need to consider how effective materials are used; what the total volume of materials used is; and not before we've done this, can we say for certain which technology imposes the greatest or the least burden on communities and natural areas. Having performed these analyses myself, I can confidently state that there's a great discrepancy between what the environmentalist movement attributes to nuclear energy in terms of mining and retrieving feedstock, and what it looks like, contrasted with virgin material requirements for any other energy source.

What about New Nuclear as a social movement?

Klein continues: "I am not talking about next generation nuclear in the same way that I'm not talking about you know real capitalism. I know that there are people, who are saying that next generation nuclear will just run off waste and we'll have no risk. I don't know about that kind of nuclear."

Why don't you *"know about that kind of nuclear?"* Is it willful ignorance? If so, you're a part of the problem. So far, you've been a champion for energy conservation and renewable energy. But it becomes increasingly more apparent that these aren't going to solve our problems. At what point are you going to reexamine your reasoning?

Klein continues: "I know about the nuclear that we have, and I know quite a bit about the risks associated with it. And we have to get away from energy models that ask other people to eat the risk."

So far, the risks associated with nuclear energy have been grossly over-estimated. And when materials usage and feedstock are concerned, nuclear has the smallest footprint of all the energy sources. Therefore, this argument is moot. We only have the macro vs micro juxtaposition. Where nuclear is a possible means to bring carbon emissions down on a macro-economic scale, while renewables are ostensibly micro-economic. But this is wrong. By definition, mass-produced technologies are macro-scale and involve large corporations by their very nature

I urge Klein to reexamine her reasoning and accept that nuclear energy must play a role in mitigating the damage from anthropogenic climate change. And that there's no valid reason to discount it. Individuals like Klein and McKibben, who I deem rational, should re-examine their reasoning, and cross the divide to create and maintain a wider and far more effective movement that will usher in a new era in energy generation and consumption. One that is based on all the low-carbon energy sources, including nuclear energy.

I want to conclude with one of Michael Shermer's remarks from the Merchants of Doubt film:

"Why don't people change their minds when new data comes out on climate? Because it isn't about the data. It's about me being a consistent tribal member and showing to my fellow tribal group members that you can count on me."

Mark Z. Jacobson — a critique

Mark Z. Jacobson tries to dictate the climate change and energy discussion. And manages to do so to some degree. He cloaks his diktats in science. Framing it in this way might get me into trouble because Jacobson's modus operandi seems to be very antagonistic. There is a great litany on the web that tells the story of how Jacobson stifles any debate. Every time a person tries to reason with him concerning his papers, and his stance against nuclear energy, he simply resorts to name-calling, ad hominem attacks and even blocking the person's communication and moving on. And when challenged academically, he lashes out. First, he tries to discredit the authors with unsubstantiated personal attacks — often based on logical fallacies. Once, the Stanford Provost for research had to be engaged simply to get Jacobson to respond to a basic criticism of one of his articles. But in the end, when his adversaries don't buckle he turns to the harshest possible response, most recently, trying to ruin the person financially and scientifically through litigation.

Case in point is the completely pointless and ugly lawsuit he filed against independent researcher Christopher Clack in 2016. We'll return to this issue later in this chapter. First, we're going to zoom in on Jacobson's work and his actions, and why I think they are immoral.

Suppose we want to end the influence of fossil fuels and try to do so using renewables exclusively. Is that going to work? That's basically the premise of Jacobson's work.

First, we must acknowledge that none of us can predict the future. Whatever an academic says about the future is probably a well-educated guess based on linear or non-linear extrapolations of our recent history. We may call these hypotheses that have to stand the test of time. And Jacobson's will be proven or disproved when the year 2050 comes to pass. That's unless you can find modelling errors, omissions, or other grievous mistakes. Then, you can disprove it straight away. We're going to examine three areas of his paper: Energy consumption projections, material requirements, and modelling errors.

Energy Consumption Projections

As humanity keeps growing, so does its need for energy. Will we be able to smooth things out? Will we find a plateau? Or will our energy demand keep rising? Or will it plummet? These are all valid questions, though hard to foresee. Therefore, I will show the scope of the problem by using the US Energy Information Agency's population and energy extrapolations.

First, we consider Table A1[29] which can be found in the 2017 International Energy Outlook by the US Energy Information Administration (EIA). For the OECD the EIA projects an annual growth rate in primary energy consumption of 0.4%, leading to an increase of 37 Quadrillion Btu* (or 37 Quads). Whereas they project an annual growth rate of 1.4% for the non-OECD, leading to an increase of 202 Quads. Let's contrast this to the population growth projections of the same regions, so that we then can boil it down to an energy per capita comparison between 2015 and roughly 30 years from now.

> * A Btu (British Thermal Unit) is a smallish unit of heat energy – a resting adult dissipates about 450 Bu/hour, or about 150 Watts – it takes several of us to warm up a room.

When we consider Table J4[30], we may note a projected annual growth rate of 0.3% for the OECD, leading to an increase of 160 million people; for the non-OECD we may note a projected annual growth rate of 0.9%. leading to an increase of 2.1 billion people. But here's the thing, this population growth is mostly centered around Africa where we will see an increase of 1.2 billion people (more than 50% of the total non-OECD population growth).

In total the primary energy consumption of mankind is projected to increase from 575 Quads in 2015 to 813 Quads by 2050. An increase of 238 Quads or 41%. Conversely the total population is projected to grow from 7.2 billion people in 2015 to 9.5 billion people by 2050. An increase of 2.2 billion people or 30%.

Btu or British thermal unit is one of the metrics we can use to appoint a value to energy. Since we're in the Quadrillions when using Btu, one uses Quads instead. Also, I prefer to use Watthours, or multitudes thereof (thousands - K, millions - M, billions - G, trillions - T), because most decarbonization pathways rely on electrification of most, if not all, of our energy production and consumption technologies. And we want to keep our metrics as uniform as possible.

We're now going to convert the energy figures from the previous page:

575 Quads → 168,000 TWh
813 Quads → 238,000 TWh

Next, we must acknowledge that electricity production is only a small part of these figures, and we must keep in mind that this is about one-third of the total primary energy input, due to the inherent inefficient nature of burning fuels for energy.

In 2015, world-wide electricity production was 23,000 TWh. At an assumed efficiency of 30% we would have to invest roughly 70,000 TWh worth of primary energy into the electric system to be able to generate 23,000 TWh of electric energy. With an installed capacity of 6383 Gigawatts we get an average capacity factor of 38%.

Jacobson assumes a 42.5% efficiency gain over our contemporary energy production and consumption infrastructure. These gains come mainly from eliminating combustion, mining, refining, and transporting fuels, and some end-efficiency gains. If that's the case then the 2015 figure of 168,000 TWh would be decreased to 71,000 TWh, and the 2050 figure of 238,000 TWh would be decreased to 101,000 TWh.

It is important to note that when you electrify all, including your generation capacity, your primary input requirement declines significantly (which also depends on additional insulation and combination measures). Suppose we create an energy mix with an average Capacity Factor of 25%, we would get the following equation:

$101,000/25 \ x \ 100/8766 = 46 \ Terawatts$

What we've done so far, is get some insight in how much power we would need to generate 101,000 TWh, if we could eliminate as much fuel inputs as possible by converting to wind and solar power predominantly. Which is exactly what Jacobson is doing. Let's consider the combination of energy resources Jacobson requires for his hypothesis to work. Below you see a simplified version of *Table 2. Number, Capacity, Footprint Area, and Spacing Area of WWS Power Plants or Devices Needed to Meet Total Annually Averaged End-Use All-Purpose Load, Summed Over 139 Countries* from Jacobson's September 6th, 2017 paper[31]. Do note that the capacity factors used come from electric power monthly page of the EIA and have all been rounded up[32].

	Capacity required (GW)	Percentage already installed	Total expected (TWh/yr)
Wind total	13,020	3.3%	39,947
Non-wind non-solar total	1,494	71.83%	5,736
PV Total	18,133	0.91%	41,328
Utility CSP	2,150	0.23%	4,335
Total	46,200	3.63%	91,346
Total Peaking/Storage	6,000	8.11%	13,636
Total All	**52,200**	**4.26%**	**104,982**

We take note of a small discrepancy of *only* 3,000 TWh between the EIA's prediction and Jacobson's total energy requirement if we could implement this higher efficiency of 42.5%. We're going to keep working with this higher efficiency, setting our bar at around 100,000 TWh of energy required per year, by 2050. The reason to stay with the TWh figure rather than the Terawatt or Gigawatt capacity figure is because the latter is dependent on your mix. It can change, as the average capacity factor of your mix can fluctuate. It's a variable, not a constant.

Two questions come to mind when I think about the future envisioned by Jacobson's theses: Does this provide grounds for a more prosperous society? How much more would we need if it doesn't?

So far, roughly 2200 Gigawatts of renewable capacity have been built (including hydro-electric and biomass). To reach Jacobson's proposed energy scenario, we must add another 50,000 Gigawatts of capacity, and do so within 32 years. If we would be able to do so in linear fashion, it would require at least 1500 Gigawatts of capacity added each year. Also note that there's a 20- to 25-year replacement interval for all the solar and wind technologies involved. This means that we must account for cumulative upkeep, adding an additional 1500 Gigawatts per year after 25 years and so on. Considering the REN21 2017 report[33], we may note that approximately 136 Gigawatts of capacity were added in 2016. To close the gap, we would have to increase today's rate of expansion of renewable resources by 1100%.

It is important to provide more perspective. In a 35-year period from 1980 until 2015 our average annual capacity additions were 110 Gigawatts per year for all electricity sources including the fossil fueled ones[34]. The highest build-rate we've achieved so far was in 2005 when we added roughly 290 Gigawatts which is still a far-cry from the 1500 Gigawatts we need. Additionally, 270 Gigawatts of these 290 was fossil fueled electricity capacity.

These figures are here to show you what we're up against. The rate of additions required for any technology to counter the use of fossil fuels is unprecedented. We've never done anything like this before. And that's not to say that we cannot do it. It is to say that it is unreasonable to put all your eggs in the renewable basket, especially in wind and solar, which bear many more problems than just having a limited expansion rate.

Material Requirements

So far, it doesn't look good. Not just for Jacobson's thesis, but for mankind overall because we aren't doing enough. Renewable additions are insufficient and are still dwarfed by the aggregate of coal and gas, especially when considered in a generation context compared to capacity additions, and this begs the question. What is the maximum capacity we can build, each year? Are we faced with any limiting factors? To examine this, we consider the permanent-magnet variant Vestas V90-3.0 (3 megawatts) wind generator, as it has the highest possible material usage efficiency[35]. And for solar panels we take the one available to us with the highest possible solar-conversion efficiency PV Panel, being the 21.5% efficient SunPower X22-360 panel rated at 345 Watts[36].

First, we're going to consider the raw figures. Remember, we're aiming for approximately 13,000 Gigawatts of Wind and 18,000 Gigawatts of Solar. Do note that the figures below are without auxiliary devices and infrastructure like mountings, transformers, switchyards, inverter, etc.

Total Wind Generator Material requirements:

$13,000/(3/1000) = 4.3$ *million wind mills*

4.3 *million wind mills* $x\ 256\ tons = 1.1$ *billion tons*

Total Solar Material requirements:

$(18,000\ x\ 1000\ x\ 1000\ x\ 1000)\ /\ 360 = 50$ *billion solar panels*

$(52$ *billion solar panels* $x\ 18\ kilos)/\ 1000 = 930$ *million tons*

The mass of all the wind generators and solar panels combined is almost twice as much of the mass of all the cars on the planet. We'd be able to build the same volume in cars in about 40 years, but don't forget that it took several decades to build all the infrastructure to build all these cars and maintain them. Just to give you an impression of the unprecedented scale involved.

By themselves, these numbers are meaningless. But from these we can deduce some interesting things. For instance, can we expect there to be a bottleneck somewhere? The key commodities I look for are always cobalt, lithium, neodymium and copper. Despite all four being available to us in relative abundance, extracting them at a rate required to realize a solar and wind powered future, might not be possible. In this case we're going to look at copper and neodymium.

2016 copper production: 19,700,000 tons[37]

2014 neodymium production: 7,300 tons[38]

We'll tackle the neodymium issue first. A 90-Megawatt powerplant consisting of Vestas V90-3.0 wind generators contains 7 tons of permanent neodymium magnets[39]. A permanent magnet contains 29% neodymium, which gives us 2 tons per 90 Megawatts of wind generators, or 22 tons Nd per Gigawatt. If we would be able to reach Jacobson's 13,000 Gigawatts of wind capacity by 2050, and do so linearly, we'd be adding roughly 406 Gigawatts of wind per year, requiring us to extract 9,000 tons of neodymium. Now here's the catch. Not all neodymium is used for the creation of permanent magnets, and not all permanent magnets are used in wind generators. According to an article, by Peiró et al, in 2013, we can break the percentages of neodymium used down in the following applications[40].

Electrical devices 58%, electric vehicles 19%, others 11%, ceramic and glass 5%, wind turbines 4%, MRI 2%, catalysts 1%.

This is an assessment of the 2010 situation. I've yet to find better and more recent information. So, this will have to do. 4% of 9,000 tons is 360 tons. And this means that by those standards we'd be able to build roughly 16 Gigawatts worth of Vestas V90-3.0 wind generators. Increases in neodymium production are required if we want to use the most efficient wind generators of all. Additions of 16 Gigawatts per year are insufficient. Remember, 406 Gigawatts per year are required. This also tells us that the current wind generator additions mostly are of a non-permanent magnet design because 54 Gigawatts were added.

Let's consider a wind generator that doesn't need permanent magnets but relies on induction generators that use copper coils to generate magnetic fields rather than having to rely on the magnetic field of a permanent magnet.

The benefit is that you don't need permanent magnets, and resource availability is not as prohibitive. We consider the Siemens Gamesa G114-2.0MW wind generator[41] and see that each wind generator relies on 0.75 tons of copper per Megawatt or 750 tons per Gigawatt. The permanent magnet units come in at 0.65 tons per Megawatt. With a requirement of 406 Gigawatts of wind capacity added per year, you'd need 300,000 tons of copper a year. Before we can reach a conclusion, we must add the analysis for solar. So far, everything looks good. 19.7 Million tons of copper are being produced, and only 300,000 tons are required.

The copper content in PV panels varies greatly. Some merely have strips of tin-coated copper across their surfaces, whereas others have copper backplates that offer higher efficiencies and strength. The SunPower X22-360 is based on such copper backplates. But I'm not going to use these in this exercise because I can't find any reliable data on the copper use in these panels. I know from SunPower's datasheet that each separate solar cell weighs 6,5 grams. The copper content is a part of that weight, but it is unsafe to assume any degree of copper in these cells. Therefore, I had to find different data. Luckily, I was able to find some data in a recent study[42] about the life cycle assessments of grid-scale PV plants.

Out of 100 Kilos of PV Panel, they expect that 110 grams of copper can be recycled. A SunPower X22-360 panel would therefore contain 20 grams of copper. Subsequently, this give us 55 tons per Gigawatt. At 560 Gigawatts (1.5 billion panels / 29 million tons) per year, the volume of copper for the panels alone would be 30,800 tons per year, which is a smidgen compared to the copper requirements for wind. But we're not there yet. We need to account for the copper in the wiring, the switching gear, the transformers, etc.

So far, the best balance of plant analysis I could find comes from SmartGreenScans[43]. At 0.8789 kilos per square meter (for the wiring alone), we'd get the following equation.

$$\frac{1.63\ square\ meters\ x\ 0.8789\ x\ 1.5\ billion}{1000} = 2.1\ million\ tons\ of\ copper$$

We then add the prior copper requirements and get approximately 2,5 million tons of copper. Which is less than 19.7 million tons. However, current additions require roughly 50,000 tons of copper, which is of course tiny compared to 2.1 million tons. And the issue here isn't that we don't produce enough copper, but that we must add another 2.5 million tons (another 12%) whereas historical the copper growth rate has never ever shown such a leap. It's either that or accept the fact that we will see continued instability in copper prices, shortages and the suchlike. Additionally, if we allocate all the extra copper to the great buildout of wind and solar, what copper are we going to use to electrify everything else and help those entering a more prosperous world do so electrically, rather than thermally. It is one thing to build the generation infrastructure required. We also must convert our existing end-use infrastructure from thermal combustion or burning to electrical conversion. Which requires great amounts of copper and aluminium.

Most 100% renewable study don't include such lines of reasoning. They just assume that there's an infinite pot of copper somewhere from which we can extract endless amounts of material at questionably high rates or maintained growth rates. And note that none of geothermal, hydro, or nuclear power have anywhere near such demands for these metals per KWh they deliver.

Modelling Errors

Rebuttals to the 100% WWS roadmap are forthcoming. First, we have the popular, non-fiction "Roadmap to Nowhere" [44], written by Mike Conley and Timothy Maloney. I could also advertise my own work called *"The Non-Solutions Project."* But I must admit that the science has moved on since I last released it. And I should update it. For I feel that it is no longer accurate. However, the mightiest blows struck against Jacobson's 100% WWS theses come from the hands of Heard et al[45] and Clack et al[46]. The first is called *"Burden of proof: A comprehensive review of the feasibility of 100% renewable-electricity systems"* And the second article is called *"Evaluation of a proposal for reliable low-cost-grid power with 100% wind, water, and solar."*

The first article (Heard et al) does not necessarily single out Jacobson's thesis, but also includes 23 other 100% renewable transition theses. The second article (Clack et al) specifically addresses Jacobson's attempt to model a future in which the United States relies solely on renewable technologies for energy generation. I'm going to focus on one very specific modelling error as it has become the first point of contention in the lawsuit of Jacobson against Clack.

The question here is: what is it? Is it a modelling error? Or an omission of a fact? Or the omission of a modelling assumption, or whatever you want to call it. Consider these simple binary statements:

Has a hydro(storage) delta of ~1.3TWh/h been modelled? Yes/No

Has a Hydro(storage) increase of ~1.3 TW been included in the thesis? Yes/No

Is it feasible to increase Hydro(storage) by ~1.3 TW of capacity? Yes/No

Has the cost for such an expansion been included in the thesis? Yes/No

The answer to the first question is yes. The answers to the corollary questions are a resounding no. When you consider Figure 4[47] of the *"Low-cost solution to the grid reliability problem with 100% penetration of intermittent wind, water, and solar for all purposes"* article, which was published in PNAS in 2015, you can clearly see that pumped hydro storage is expected to sustain generation of electricity for approximately 1.3 TWh per hour for at least half a day.

This is in direct opposition to the assumed maximum charge (Discharge) rate of PHS (Pumped Hydro Storage) of 57.78 Gigawatts and the 0.808 TWh of assumed energy storage capacity or total Hydropower capacity of 87.48 GW as listed in Tables S1 and S2 of the appendix[48] to that same article. and it is therefore fair to conclude that Jacobson must have A. Made a modelling error, or B. Omitted essential facts in his thesis to maintain these modelling outcomes, or C. Made unsound, unwarranted and unsubstantiated assumptions. Which of it is, I leave up to you, as making any statement in this regard seems to be interpreted as defamatory or libelous according to Jacobson and his attorneys.

The Lawsuit

Which brings us to the final part of the commentary on Jacobson's work in anti-nuclear activism.

The uncovering of Jacobson's flawed work has sparked a debate that is about science, but Jacobson doesn't want it to be about science. He thinks that the evaluation paper written by PNAS damaged his credibility, as if he had any to begin with. But Clack's paper argues that there are modelling errors and unwarranted assumptions in Jacobson's thesis. After its publication, Jacobson tried to get it retracted. A dispute between Clack, PNAS, and Jacobson ensued, ultimately leading to Jacobson filing a libel lawsuit against Clack and PNAS. What is more, Jacobson claimed ten million US Dollars in damages. Effectively threatening the only independent author of the evaluation paper with financial ruin.

Through these actions, Jacobson has effectively taken the scientific method hostage. And this was later confirmed by Ken Caldeira who wrote this on behalf of someone else[49]:

"The litigious activities of Mark Z. Jacobson (hereafter, MZJ) have made people wary of openly criticizing his work.

I was sent a PowerPoint presentation looking into the claims of Jacobson et al. (PNAS, 2015) with respect to this hydropower question, but the sender was fearful of retribution should this be published with full attribution. I said I would take the work and edit it to my liking and publish it here as a blog post, if the primary author would agree. The primary author wishes to remain anonymous."

While Jacobson set out to prove that something can be done, honest scientific inquiry would assume that something cannot be done, unless certain criteria are met. Testing hypotheses through rigorous well-substantiated modelling and experiments is at the heart of academic progress. Once you publish your hypothesis, with all its test results, it is up for scrutiny. And when other's find mistakes in your work, despite peer-review, you should re-evaluate your hypothesis, correct for the errors and retest or remodel. That's how we ensure that our findings are sound. So far, Jacobson has only disputed the critiques of his work.

Jacobson and his co-authors have been prolific in creating new papers which depend on the consistency and validity of prior papers they use to substantiate their new work. Earlier in February 2018, Jacobson and others published a child-paper in the Renewable Energy Journal of February 2018. The question is whether this one can stand as the disputed paper is referenced as well. Additionally, given the fact that Jacobson is willing to take you to court when you publish a critique of his work, it is conceivable that academics will shy away from evaluating this new paper and addressing fault if found. Thus, stifling rigorous academic discourse.

A fundamental question that comes to mind is this: What if any serious flaws like the "hydro-argument" are uncovered in other theses of Jacobson. Would it then cause his entire work to collapse? Since he references prior work to substantiate new claims, retracting one of these would cause a cascade of scientific articles to become unsound.

Questionable Integrity

I would not go here if there were no inconvertible proof, but clearly Jacobson's ego is so easily hurt that he must lash out at every possible occasion. He has blocked people who tried to reason with him. And he has vilified those who have refuted his arguments, up to the point of trying to ruin those who dared to speak up against him. But this story isn't over. On February 18th, 2018, a group of people tried to petition Jacobson to withdraw his lawsuit and to settle his dispute with Clack and PNAS where it should, in academia. The petition text (which I partially drafted):

"[The] Don't Sue Science Petition

In 2015, Mark Z. Jacobson, a Stanford University professor, published a peer-reviewed article in a scientific paper called the Proceedings of the National Academy of Sciences (PNAS). In this article, it is claimed that 139 countries can decarbonize their economies completely by transitioning to an energy system that is powered by renewable sources exclusively and do so at reasonable cost.

Though the article was heralded by many as solid proof that we can transition away from fossil fuels by deploying renewables, experts in climate change and energy fields were not convinced. Multiple books and peer-reviewed articles have been written calling Jacobson's claims and methodology into question, including Christopher Clack et. al's "Evaluation of a proposal for reliable low-cost grid power with 100% wind, water, and solar", peer-reviewed by twenty-one content area experts and published in the Proceedings of the National Academy of Sciences.

*Rather than responding to these critiques **with corrections**, Jacobson is suing both lead author Christopher Clack and the National Academy of Sciences for $10 million in damages. That's where we draw the line. If this lawsuit is allowed to stand, it will have a chilling effect on any researcher who dares question the work of another published author. For the sake of continued scientific progress,*

We, the undersigned, stand united in defense of

Honest scientific inquiry

Rigorous academic discourse

Settling scientific disputes with research, not legal filings

We demand Mark Z. Jacobson drop his lawsuit. #dontsuescience."

A day after, people were noticing that the petition was no longer available. When one of the organizers inquired as to why the petition was pulled from the website, the reasons stated where:

"Upload a petition about any issue that is currently before the courts, or will or could soon be before the courts, or use a petition as an instrument to engage in pre-litigation or litigation positioning. Petitions must not discuss evidence before the courts or show contempt of any court in any jurisdiction. A petition may not lobby court official in any matter that is before the courts. A petition cannot lobby jurors or make any comment

about them. A petition should not, as a general rule, lobby state or federal prosecutors   or   judges.

Also, we received "Report Abuse" from Mark Z. Jacobson."

It wasn't at all surprising that Jacobson reported the petition as abuse. At some point people should wake up to the reality that his self-righteous pleas are nothing but attempts to defend his fragile ego and his opinions at all cost. As if matters couldn't get any worse, he resorted to name-calling and tweeted the following on February 19th, 2018.

"Nuclear trolls have their defamatory petition removed due to abuse. Amazing the lengths haters [twitter handles used] Ben Heard, Matthew Herald, Eric Meijer and others go to smear. Their claim that the suit is about science is completely wrong. It is about false facts."

And he caps it off with this one, in which I am mentioned as well! What an honor…

"And this creepy stalker Mathijs Beckers admits to being one of the organizers of the petition, and copies Ken Caldeira and PNAS news and Michael Shellenberger in his posts. Obviously trying to influence the lawsuit and spread hate at the same time. Disgusting."

Jacobson was completely right, we were trying to influence the lawsuit. In fact, we were trying to reason with *him* to end this lawsuit. Also, there's no law against expressing your concerns, even while a lawsuit is going on. In fact, no injunctions were issued against it, nor will there ever be. So, we're free to say whatever we want thanks to the first amendment (even though it doesn't apply to me because I am a Dutch citizen—this kind of freedom is enshrined in Article 7 in our Constitution). If something could be regarded as disgusting, it is his vile attempt to ruin a person financially because he wrote an honest and accurate critique of Jacobson's work, trying to mask it as a libel suit *("It is not about science")*. As if he couldn't stoop any lower, he ends his inane tweet with a single worded sentence. A Trumpism?

On February 22nd, 2018 Jacobson decided to withdraw from the lawsuit. Which can be regarded as a victory for justice. But it wasn't a true victory…

Jacobson writes: *"Yes. Not only did we request corrections of factually false statements and/or a retraction before the lawsuit to avoid the lawsuit entirely, but we also offered to drop the lawsuit entirely if PNAS would publish the following simple factual corrections."*

And therein lies the truth. He maintains that Clack's paper contains false statements and that Clack's paper must be amended or retracted. Clack didn't acknowledge that these were false statements (which they weren't as far as I am concerned). Jacobson also offered to amend his own paper to account for his mistakes pointed out by Clack. Which begs the question. Was Clack right after all? In my mind, there's no doubt that Clack made correct assessments about Jacobson's paper. Jacobson appears to agree, manipulatively so.

When asked by Stanford University Professor Ken Caldeira if Jacobson would offer *"to reimburse legal expenses you have caused people to incur? Your lawsuit has caused real injury, costing people money and wasting their time."*

Jacobson, yet again, showed his true colors: *"Ken, you still don't get it. It is YOU and your 20 coauthors who recklessly published false facts in three areas and refused to correct them. You have no-one to blame but yourself. Take responsibility and stop blaming others for your own actions."*

Naturally deflecting using uppercase wording and scathing accusations, which seems regular practice when Jacobson addresses his *"antagonists"* (emphasis added) and proclaims to see *"false facts"*—Jacobson seems unaware of what "fact" means.

It seems that this lawsuit was nothing but a feeble attempt to hide Jacobson's mistakes, but this came at great cost for those who honestly tried to evaluate his thesis and uncovered the mistakes in the first place. These actions by Jacobson erode Stanford University's reputation. There can be some debate about the nature of the mistakes. Were they intentional? Did Jacobson hide assumptions? For what reason? But To me, his academic credibility has committed suicide several years ago. Being unable to cope with any kind of criticism and his unfaltering insistence on being right are the main weaknesses that invalidate his standing as a scientist. Also, his Wikipedia page looks more like an autobiography than anything else. Not so surprising given the fact that a user named *"Mark Z. Jacobson"* has contributed 37% of all the content[50]. It seems that everything he touches is tainted by his own confirmation bias.

So far, there are no signs that he will accept that any of his models and assertions are flawed. Still, he will reap the rewards of his fame and popularity. In fact, the Massachusetts Institute of Technology (MIT) invited him as a keynote speaker for the MIT Energy Conference in 2018 (MITEC2018). And this only served to perpetuate his overzealous dogmatic anti-nuclear / 100% renewable narrative, and thus maintains the divide between those who want to fight climate change with whatever means necessary, and those who say they want to fight climate change but believe that nuclear energy is unnecessary.

Does Jacobson have the capacity of self-reflection? I guess that's shooting for the moon. A man whose grandiose plans have led to friendships with and support from television personalities and other celebrities is unlikely to give all of that up.

To state that *we who oppose him* are *"trolls"* and spread *"hate"* (his words) is by definition wrong. We are concerned scientists, engineers and intelligent people who see flaws in his reasoning and try to address them in a sensible way. It is Jacobson, who consistently resorts to logical fallacies, childish ad hominem attacks and misrepresentations. We don't want people to hate him (at least I know I don't). On the contrary, we want people to see that he has been wrong and tell him that he is (as are we all) fallible and must re-examine his reasoning or reveal his own bias.

Naomi Oreskes

As a former Geology student, and now professor of the History of Science, at Harvard, no less, Naomi Oreskes must have known better than to choose to side with scientists whose popularity seems more important than the diligence with which their theses are substantiated. Case in point, subject of the prior chapter: Mark Z. Jacobson. The Hubris surrounding Jacobson has caused scientists, politicians and stars to gravitate towards him. Including Oreskes. However, by excluding nuclear energy, he envisions a longer, less effective, and ultimately unattainable decarbonization pathway. And therefore, the people supporting him should come to their senses, and see this exclusivism for what it really is, a dangerous climate gamble, with actual repercussions for people and nature all over the globe.

On December 16th, 2015 The Guardian published an article[51] written by Oreskes called *"There is a new form of climate denialism to look out for – so don't celebrate yet"* From which I am going to quote.

Oreskes writes: There is a new, strange form of denial that has appeared on the landscape of late, one that says that renewable sources can't meet our energy needs."

This argument is the basis for this book. And I think that Oreskes easily brushes over the fact that in this specific case, the science, indeed, isn't settled. And why is that? Because we're building energy models, trying to project into the future, based on unsound assumptions. Additionally, Oreskes makes it look as if there is a false equivalence between the idea that we can push the brakes on temperature increase by deploying renewables exclusively, and the counter argument that it is probably not going to happen by deploying renewables exclusively. The latter, in her eyes, ostensibly being wrong. And we will come to this later.

Oreskes writes: "Oddly, some of these voices include climate scientists, who insist that we must now turn to wholesale expansion of nuclear power. Just this past week, as negotiators were closing in on the Paris agreement, four climate scientists held an off-site session insisting that the only way we can solve the coupled climate/energy problem is with a massive and immediate expansion of nuclear power. More than that, they are blaming environmentalists, suggesting that the opposition to nuclear power stands between all of us and a two-degree world."

This passage contains two issues. The first being that there are Climate Scientists (Tom Wigley, Kerry Emanuel, Ken Caldeira, and James Hansen)[52] who want to educate the public about the necessity of including nuclear to help shape policy regarding climate change and energy. All of this happened while there was a critical, yet failed, climate summit in Paris. In fact, the scientists came out afterward, taking notice of an ineffective solution, despite the widespread hubris that ensued after a unanimous conclusion at COP21. In fact, many individuals would come out and claim COP21 to be a great success.

Secondly, the accusation that the climate Scientists blame the environmentalists who are in opposition to nuclear power for the possibility of us punching through the 2 degrees threshold. Actually, Brown and Caldeira[3] have made a compelling case for this happening. In fact, they predict that we will punch through a 4 degrees threshold if we keep going about our business without really reducing our greenhouse gas emissions. And so far, nothing has happened to suggest that we will be able to cut these emissions. And it is the mutual finger pointing that contributes to this stalemate.

In effect, the aforementioned scientists are right. Years of anti-nuclear activism have not only caused serious stagnation in nuclear development and deployment, but also fostered a deep seated anti-nuclear sentiment. Perhaps even amongst those addressed in this chapter. These sentiments are represented across the board. From average people on the streets to policymakers in office. And this is a recipe for unnecessary stagnation in addressing climate change. As we have seen in the US, UK, Finland, and France. Think about the AP1000 and the EPR which were mired in

41

bureaucratic indecisiveness and constantly changing demands, well after the first blueprint was approved and paid for and first ground was broken.

*Oreskes writes: "That [the need for nuclear] would have troubling consequences for climate change if it were true, but it is not. Numerous high-quality studies, including **one recently published by Mark Jacobson of Stanford University**, show that this isn't so. We can transition to a decarbonized economy without expanded nuclear power, by focusing on wind, water and solar, coupled with grid integration, energy efficiency and demand management. In fact, our best studies show that we can do it faster, and more cheaply."*

These studies are flawed. Most of the all-renewable models assume a status-quo between OECD and non-OECD per capita energy consumption. Additionally, all of them model decreasing energy demands, which is a direct negation of the fact that emerging countries are trying to increase their energy supply and that these are the countries which have growing populations and are contributing to the net addition of two billion people by the end of this century. Facilitating even the most basic needs for these people is going to require a major fraction of the energy that we use today.

Also, these renewable transition studies disregard essential resource limitations and storage complexities. In fact, Jacobson's thesis (for the US), which is often touted as the most credible one and therefore enjoys a lot of popularity, relies on 1300 Gigawatts of Hydro Capacity for the US alone. An increase from 85 Gigawatts today. Needless to say, the United States will never build 1300 Gigawatts of Hydro Capacity[53], nor will they ever be able to expand extant capacity by adding generators. It simply is not possible. The laws of physics are prohibitive in this regard. Nevertheless, Jacobson has modelled 1300 Gigawatts of hydro (pumped storage) capacity but didn't specify that amount of hydro capacity in his table of future technology requirements.

Oreskes wrote the previous quoted section more than a year ago and so far, I've not seen any evidence to suggest that she has changed her mind on this issue.

Oreskes writes: "The reason is simple: experience shows that nuclear power is slow to build, expensive to run and carries the spectre of catastrophic risk. It requires technical expertise and organization that is lacking in many parts of the developing world (and in some part of the developed world as well). As one of my scientific colleagues once put it, nuclear power is an extraordinarily elaborate and expensive way to boil water."

Each of these points will be addressed in the final chapter of part III. Suffice to say that some of these criticisms are valid to some extent, but most are not.

Oreskes writes: "The only country in the world that has ever produced the lion's share of its electricity from nuclear is France, and they've done it in a fully nationalized industry – a model that is unlikely to be transferable to the US, particularly in our current political climate."

In this passage Oreskes makes a credible argument in favor for Nuclear Energy. Just because France did it in a nationalized context, doesn't diminish the fact that they've done it, and done so in a short period as well. If we consider the French nuclear buildout, we may note that they have built 90% of their capacity in a timeframe of 21 years[54]. We may also note that the French built Blayais 1 in 4.9 years and that they finished 90% of their reactors in 6.7 years from start to finish.

Therefore, it is justified to assert that Oreskes has come to the wrong conclusion. It doesn't teach us that it is impossible, because it was a nationalized effort and is hard to transfer to the US. It teaches us that when the framework around Nuclear deployment is set up correctly, and this doesn't necessarily mean institutionalized, we can deploy nuclear power plants rather quickly, and there's ample proof of other countries who are efficient nuclear builders as well. Consider for instance South Korea, Japan, and Switzerland. (*Later, in this book we will discover the success factors for deploying nuclear fast and comparatively cheap.*)

In effect, it is exact the same argument with renewables. If only the correct prerequisites are met, the technology will be deployed quickly. And do note that I encourage this, but that it is paramount not to exclude nuclear.

Oreskes continues: "Even in the US, where nuclear power is generated in the private sector, it has been hugely subsidized by the federal government, which invested billions in its development to prove that the destructive power unleashed at Hiroshima and Nagasaki could be put to good use."

Could be, and has been[55], and will be. Note the reference to nuclear warfare, as if that is an argument against civilian nuclear energy. This is a moot point because the days of breeding weapons-grade material in reactors are over. In fact, nuclear power reactors have contributed to the reduced stockpile of weapons of mass destruction through the Megatons to Megawatts program[56]. We can use weapons grade material as fuel in just about any reactor, even the old ones. Additionally, frequent oversight is there to ensure proliferation doesn't happen. Also, making Plutonium for weapons requires a very peculiar fueling and refueling cycle which can be picked up by satellites and other technologies thanks to shifts in thermal- and power output. To conflate nuclear energy with nuclear weapons is like trying to ban pears because they contain formaldehyde[57].

Oreskes writes: "The government also indemnified the industry from accidents and took on the task of waste disposal – a task it has yet to complete."

Oreskes is probably speaking about Yucca Mountain, a repository built by the US government to store spent nuclear fuel. A decision with which I strongly disagree. Why would we want to sequester spent nuclear fuel when we can store it effortlessly in renewable and robust dry-cask storage containers? Intermediate storage and continuous management should be pursued. Simply because this means that we can retrieve the nuclear materials, which can be put to good use. Just because the fuel has outlived its purpose in a Light Water Reactor, doesn't mean that it is now completely useless. In fact, only a fraction of the energy has been extracted. And there are many possible applications for fission products. Consider spent nuclear fuel to be a left-over from the Nuclear Bronze Age. We're about to transcend into the Iron Age of nuclear power, where liquid fueled reactors can address almost all, if not all the issues raised by antagonists to nuclear energy—remember, however, that current reactors are fine!

Oreskes writes: "So far no one has proposed a plan to do that, and we probably won't get very far if the alternatives to fossil fuel – such as renewable energy – are disparaged by a new generation of myths."

Is Oreskes paying attention? Or is she quoting these people out of context? When we consider James Hansen, Ken Caldeira, Thom Wigley, and Kerry Emmanuel, and look at their motivation to address the energy issue from a more nuanced perspective, we may take note of their sincerity and genuine sense of urgency. Let me show you by quoting them directly.

Ken Caldeira kicks off: "years ago I was protesting against nuclear power at the Shoreham nuclear power plant on island and I was arrested for protesting nuclear power..."

Caldeira continues: "it's not about either-or, we're not talking about favoring solar power, wind or nuclear power. I'm in favor of anything that can prevent climate change and preserve the environment and allow for people to get food and health care and education. I guess the basic plea here is: let's focus on the climate agenda and the climate agenda is about supplying energy in a way that does not damage our environment and we need to allow technologies to compete on their own merits."

Kerry Emanuel takes over: "But that study of the climate system has very strongly led us to the conclusion that we are inferring unacceptable risk for a future generation."

Emanuel continues: "Why are four scientists, who don't have strong backgrounds in nuclear physics, here talking to you today about it nuclear energy? It's because we're scientists, and we can do the math, right?

If we truly are sincere about solving this problem, unless the miracle occurs, we are going to have to ramp up nuclear energy very fast. that's the reality that's not my ideology I don't care whether its nuclear. Like my friend Ken said, we don't care whether it's nuclear, solar or hydro. Whatever combination works. The numbers don't add up unless we put nuclear power in the mix."

Tom Wigley takes over: "We're not promoting nuclear, we're promoting a level playing field."

James Hansen speaks: "I like to emphasize the climate impacts that are irreversible. We are at a point, now, where it's extremely dangerous. We are at the point where if planet gets much warmer we are going to get instability of ice sheets and sea level rise of at least several meters, and the consequences of that are almost incalculable. half of the large cities in the world are on coastlines."

Hansen Continues: "The other thing that's irreversible is extermination of species. If we stay on business as usual, the IPCC estimates that, by the end of the century, we could commit a quarter to a half of the species on the planet to extinction

Hansen Concludes: "We know that using fossil fuels is not safe. It is very dangerous, and we have to face the fact that this danger of fossil fuels is staring us in the face. It's absolutely a hundred percent certain that we've got a very dangerous situation, and for us to say oh we're not going to use all the tools that we, to try to solve it, is crazy. we have to we have to use all of the things that we have at our disposal and clearly nuclear power, next generation nuclear power, especially, has tremendous potential to be a big part of the solution."

Is it fair to conclude that Oreskes is mislabeling genuine concerns as myths? So far, these scientists have tried to determine whether renewables alone would be sufficient to quickly decarbonize our energy system and found that including nuclear in a future energy mix will augment decarbonization pathways significantly. Their concerns are shared by many other scientists. I don't understand why Oreskes would call these clearly enunciated and substantiated concerns, myths. It seems to be there to pit the reader against these scientists, and undermine their credibility, which is far from laudable.

Oreskes writes: "If we want to see real solutions implemented, we need to be on the lookout for this new form of denial."

It surprises me that she didn't call these scientists out on their expertise. Clearly their expertise in climate change science warrants a sidestep towards looking for credible solutions. And when solutions are proposed, and seem too optimistic, their concerns should certainly be noted, and not dismissed. It is rather confounding to see Oreskes turn on one of her former allies, James Hansen. I truly hope that they're still on speaking terms, and that Oreskes comes to her senses.

Oreskes writes: "The key to decarbonizing our economy is to build a new energy system that does not rely on carbon-based fuels. Scientific studies show that that can be done, it can be done soon, and it does not require nuclear power."

The second part of the sentence (*"and it does not require nuclear power"*) is puzzling as it doesn't convey the correct sense of urgency. How urgent is our problem? And how quick can we solve it, with all our current, and future means? Rather than: Can we do it with this suite of technologies, exclusively? Disregarding nuclear in a potential decarbonization mix only serves to reduce the speed with which we can defeat the influence of fossil fuels and with it the driving force behind anthropogenic climate change. Remember, we have about 600~700 Gigatons of CO_2 left to emit before we push over the 1.5 degrees threshold, and we're emitting about 35 Gigatons per year. Time is running out…

According to Jacobson we could electrify almost all energy consumption by 2050. However, I've shown you that his plans are based on unrealistic assumptions. Also, there's no compelling reason to stop building nuclear reactors or to shut existing ones down. There is no reason to assume that building a nuclear reactor cuts into the budget for renewables. As far as I can tell, renewables are being subsidized quite vigorously everywhere, not just in the US[58]. In fact, if we consider Germany, for instance, we may note that they have pumped roughly 220 Billion US Dollars into renewable energy[59]. And based on my own research, I concluded that said amount of money could've been spent better and would have enabled the Germans to close approximately 63% of their coal and gas and biomass burning capacity. Thus, greatly improving their carbon footprint. And truly making them the climate leaders some people mistakenly think they are.

Let it be stated for the record, that I, like Oreskes, dislike organizations like the Cato Institute, the Heritage Foundation, and the American Enterprise Institute. And to cap it off, just in case people doubt my motivation. I am a progressive, semi-socialist. No one can claim that I am aligned with people who want to break down social security and shrink the government. On the contrary, I think the government should do more for its citizens. Think about providing affordable single-payer healthcare; Basic Income; building an excellent and accessible-for-everyone education system; Public Transport; Spearheading innovation in technology; and so on. I tell you this to show you that people can come to the same conclusions in terms of science, regardless of their world-views. Energy should not be a split-ideological issue. Just as Oreskes knows that climate change science shouldn't be.

Despite my critique on her year-old article, I wish to reach out to Naomi, and tell her that there's a lot of us who are on her side with almost 95% of the issues, and that there are valid scientific reasons to require renewables and nuclear energy to work together. It is not constructive to alienate scientists who enunciate this clearly by calling them *"a new kind of denier"* or as Jacobson once stated, *"having extremist views"*.

We don't have fifty years…

The United Nations

There are now annual climate conferences which are organized under the flag of the UN. We call these COP's, which is an abbreviation of the impossibly long *"Conference of the Parties to the United Nations Framework Convention on Climate Change."* The last part is abbreviated UNFCCC.

At these conferences, leaders from all over the world try to reach agreements on how to move forward on climate change. So far there were 23 COP's. The last was in Bonn, Germany, in 2017. Ostensibly, the most important agreements, so far, have been made in Kyoto, in 1997 and in Paris, 2015. In Kyoto the convening countries agreed to reduce their carbon dioxide emissions to 18% below 1990 levels. And in Paris they agreed to keep the average temperature from rising another 2 degrees. These agreements have not been effective. In fact, the opposite is true. Carbon emissions and fossil fuel consumption keep rising with no end in sight.

Consider what James Hansen had to say during an interview[60] during COP23, in Bonn, at the end of 2017 (slightly redacted):

"I would say there's very little progress because there are no reductions in global emissions of carbon dioxide, and if you look at the amount of CO_2 and methane in the atmosphere, it's actually growing more rapidly than it was two years ago.

I thought what would happen as a result of the Paris agreement was very little. It's analogous to the Kyoto Protocol, you know. Their politicians agreed that climate was a problem and nations would try to reduce their emissions. In fact, emissions accelerated, and the rate of growth increased. If you look and if you look at developed countries their emissions peaked in 1980 and since then have been flat. There's no evidence of an impact of the Kyoto Protocol or the Framework Convention in the 1990s.

Now, we got to 2015 and we have the Paris Protocol. All the politicians clapping each other on the back as if something had been accomplished, but there's not going to be any reduction in fossil fuel use as long as fossil fuels are the cheapest energy, and that's the situation.

We have to make fossil fuels include the cost to society. That means the air pollution cost; the water pollution cost; the climate change cost. So, we have to add a carbon fee or a carbon tax which has to be across the board. Oil, gas, and coal. Not some cap and trade gimmick, which does almost nothing. The politicians have not taken the needed actions. So, in that sense there, there's been no progress."

The relevance in this chapter comes from the fact that the COP meetings have had little to no effect. That's unsurprising because it seems that the focus is on the wrong issues. What are they doing at the COPs? Our leaders are putting up a show, talking about cap and trade, reducing emissions, some people want to make this a gathering about renewables and energy conservation, while the actual problems remain unaddressed. It is all well and good to talk about reducing emissions but if no solid proposals are made that facilitate these changes, they won't happen. And that's symptomatic for all these international gatherings. People acknowledge the problem, but they don't commit to actual solutions. James Hansen is correct, and his messages should be considered more carefully. Instead of making vague promises, world-leaders should commit to actual solutions. It might seem technocratic of me, but our suite of solutions has been known since the early 60's, we only need to commit to all of these technologies. We have nuclear, hydro, geothermal, wind and solar. The nuclear solution has been known to the highest authorities since 1962. For various reasons we fail to commit.

Alongside the COPs one of the branches of the UN is organizing *"Sustainable Innovation Forums"* (SIF). The UN Environmental Programme (UNEP) which is led by director Erik Solheim is co-organizer of these SIFs. Here's UNEP's mission statement as excerpted from their webpage.

"The United Nations Environment Programme (UN Environment) is the leading global environmental authority that sets the global environmental agenda, promotes the coherent implementation of the environmental

dimension of sustainable development within the United Nations system, and serves as an authoritative advocate for the global environment.

Our mission is to provide leadership and encourage partnership in caring for the environment by inspiring, informing, and enabling nations and peoples to improve their quality of life without compromising that of future generations."

Additionally, I excerpted this from their website:

"Energy drives economies and sustains societies. Energy production and use is also the single biggest contributor to global warming. The energy sector accounts for about two-thirds of global greenhouse gas emissions attributed to human activity.

More than a billion people still lack access to electricity, while 3 billion rely on dirty fuels like charcoal and animal waste for cooking and heating. Our challenge is to reduce our reliance on fossil fuels to produce electricity and heat and power our transportation systems, while making reliable, clean and affordable energy available to everyone on the planet.

At UN Environment, we believe that sustainable energy presents an opportunity to transform lives and economies while safeguarding the planet. That's why we're working with governments to help them improve energy efficiency and increase the use of renewables in their countries and cities. We aim for sustainable energy to lay the foundation for resilient, low-emission economies and societies around the world."

Ruining the plan: that last paragraph ruins the entire thing: *"We're working with governments to help them improve energy efficiency and increase the use of renewables in their countries and cities."* It is not about deploying renewables. It is about deploying energy sources that help us defeat carbon emissions while simultaneously helping the people progress. You so passionately advocated earlier to develop a modern, secure, and prosperous way of living. what you call renewables require a great deal of support in order to be part of a reliable system.

So far, I've not seen any concrete actions from UNEP that move things forward on the climate change agenda. What about being a champion for actual solutions to emissions from fossil fuels? So far, I've seen nothing but hollow promises from UNEP. Consider this quote from Solheim: *"We need to prove that **protecting the environment is profitable** and in everyone's best interests. We can do this by holding up **successful examples.**"*

"Everyone's best interests" might not include the survival of the biosphere of planet earth? Which begs the question: what successful examples? The successful implementation of some futuristic technology in and of itself is not a successful example. The only measure of success in our current state of emergency is our ability to stop the expansion of fossil fuel usage, the lowering of annual carbon emissions, and the overall improvement of health and safety for everyone on the planet. So far, France, Ontario and a few others have shown us good examples. Otherwise, it isn't looking good. And, there's an odd censoring of nuclear successes by bureaucracies, including UNEP. So, there's no cause for optimism, and there are no successful examples to hold up. Or they must mean those insignificant countries who managed to get their energy from wind and solar (discounting their oil- and gas imports), but whose populations use less energy than one of the boroughs of New York uses in a month, in some cases, even a day.

As you've been able to read in previous chapters, the term "renewable" has become synonymous with anti-nuclear dogma. For instance, we may consider Erik Solheim's tweets:

"Offshore wind energy in UK will be cheaper than electricity from new nuclear power for the first time! A milestone!"

"US utility dumps nuclear plant, will invest $6bn in solar and batteries. Solar beating out 'uncompetitive' nuclear!"

"Fascinating! New solar plant will be built near site of Chernobyl nuclear disaster."

"It's decided! Switzerland says no more nuclear. Path forward is renewables!"

One might be mistaken, but it looks like Solheim really doesn't like nuclear energy and is rooting for renewables. I question whether he has his priorities in order. Shouldn't there be a reference to coal and gas instead of nuclear in front of every exclamation mark? What is he trying to say? That we need to defeat nuclear energy in order to mitigate the damage from climate/ocean change, and help billions of people achieve a sensible level of prosperity? At this writing, Switzerland is restarting an old nuclear plant explicitly to reduce pollution. China is staring new ones. Others around the world are being similarly responsible to our descendants, as JFK hoped back in the sixties.

What we need is to turn Solheim's narrative on its head. Though it is true that we must decarbonize, to reach deep decarbonization we need all low-carbon energy sources. The term renewable is being abused. For instance, do you consider the use of trees for energy renewable? It is. But it is far from low-carbon and carries with it a heavy burden on the biosphere due to the loss of mature trees and habitat for wild animals. *Renewable* shouldn't be the currency used to enact a future for mankind, it should be energy prosperity based on whatever low-carbon energy source we can get our hands on.

Low Carbon Energy from Nuclear: The World Nuclear Association (WNA) is working toward prosperity with low-carbon energy. The WNA mission: *"is to promote a wider understanding of nuclear energy among key international influencers by producing authoritative information, developing common industry positions, and contributing to the energy debate."* Naturally, the WNA wants a seat at the table at high-profile events like COP and the Sustainable Innovation Forum. In fact, they should be there. And it's not even as if they are asking for a free lunch. The WNA offered to become one of the key sponsors of the Sustainable Innovation Forum of 2017 in Bonn. And for a long while it looked as if the WNA was accepted as a sponsor.

What if I told you that UNEP eventually snubbed the potential sponsorship from the WNA? Surely it wouldn't surprise you, given the fact that their director has been enunciating clearly that he has been happy about renewables being able to beat nuclear at some point in time, exposing a willful ignorance of both science and energy. What ensued was a concerted effort of several advocates, including myself, to get UNEP to accept the sponsorship proposal by the WNA. However, they did not budge, and the Sustainable Innovation Forum eventually continued without any involvement from the WNA. That is not to say that there weren't any voices in favor of nuclear at the conference. Generation Atomic, Energy for Humanity and Bright New World amongst others were present in Bonn to advance an agenda of inclusion for nuclear technologies in the name of addressing climate change.

Eric Meijer, who is program director at Generation Atomic, a grassroots nuclear advocacy group, was able to speak to Solheim, and what he said was somewhat puzzling, though telling.

"I got a chance to talk to United Nations environmental program director Eric Solheim. He was a friendly guy, we had a great conversation. but he made it clear that he was more concerned about the political implications of being seen as pro-nuclear than about bringing the bottom billion out of poverty and curbing the effects of climate change in future generations."

And that's where we are. People in leadership at important organizations are afraid of losing face. They are afraid of the political repercussions if they are seen endorsing nuclear power in any way, shape, or form—remember Bill McKibben? This has to end. Or, agencies like UNEP must themselves pay a carbon tax.

Here's what James Hansen had to say about this:

"Erik [Solheim], you know, I've met you before and we're very much on the same page with regard to climate change, but boy if we don't allow nuclear a fair seat at the table I just don't see how we're going to solve this problem on the timetable that it has to be solved."

And that's the whole issue. That's why I write these books. The timetable involved here is far too short to allow for an all-exclusive buildout of renewables. In fact, research now shows the timetable shortening dramatically[61]. In the meanwhile, we're up against organizations and individuals who cling to the unrealistic hope that renewables might be able to satiate the great demand for energy of future generations while simultaneously getting rid of fossil fuels, perhaps even against their better judgment. However, mathematics and physics aren't on their side. The fate of billions of people and the wellbeing of our biosphere hang in the balance. Naive political expediency regarding renewables should be ditched in favor of rigorous adherence to scientifically valid and practical solutions. Wind and solar and a couple of other technologies are additions, not cornerstones to a clean-energy future, especially given the fact that the nuclear industry is poised to revolutionize itself with new construction and rapid development of Generation IV technologies which bring modularization, higher fuel efficiencies, waste-reduction, and safety innovations to the table. Nuclear power should, therefore, be an integral part of the movement to defeat carbon emissions. UNEP's current anti-nuclear bias foolishly endangers our planet and all its passengers. President Taft said: *"Politics is the art of the possible"*. What Taft left out was: *"short term"*. UNEP owes us all engagement in long-term politics, based on facts.

UNEP can take a responsible step by inviting the World Nuclear Association to sponsor and join equally in all upcoming high-level conversations.

Part two: Reality Check

Population growth & Energy Reality

When we consider the teachings on demographics from the late Professor Hans Rösling, we learn that the world's population will keep growing and might end up somewhere between 9 and 11 billion. With 11 billion individuals being a maximum value. In terms of demography and growth the populations of Europe, the Americas, and Asia have settled down and become somewhat stable. This means that when you consider the spread of the population over different age groups ranging from newborns up to the elderly, and consider the death and birth rates, the number of people in a specific region remains relatively stable.

In fact, Rösling postulated that the population of Asia will decline and stabilize again. But that doesn't mean that world-wide population growth will stop altogether. The decline of Asia will be picked up by Africa. Africa's population growth is still picking up in speed. This is because less children die due to improved hygiene and better access to fresh water, food, medicine and services. But also, because Africans becoming older on average, which means that the number of people increases as less people die before reaching old age. But also, the number of babies born per parent-pair are still higher than two on average. So, we will get a fill-up of adults. Eventually and probably, the fertility rate in Africa will stabilize and we will have an average of 2 children per parent-pair.

This world's population will grow with an additional 3 to 4 billion from now until 2100. Basically, it means that mankind will mature by 2100. By then the age-demographics of our species will be spread out quite evenly across the planet. I think this is a good sign. Even though it will make our climate change problem bigger, it is a sign that we as a species are becoming healthier and wiser.

I didn't see it this way earlier, but Rösling's explanation makes sense. The best lectures he gave, available online, can be found on the *"Think Global School"* YouTube Channel.

Curbing population growth is starting to become unnecessary, because we already have had the peak in child births[62], so that's a problem we don't need to worry about. What we must do, however, is make sure that those 4 billion people, plus the 10 percent living in extreme poverty today[63], get access to clean (running) water, food, hygiene, medicine, and excellent essential services for health, safety and education. But most important of all, to give young people the chance to develop their own talents and fulfill their own dreams.

If you want to know how well our energy is used, you can consider energy flow diagrams. The Lawrence Livermore National Laboratory creates very good energy flow diagrams for the US[64]. When you consider the 2016 version you may note that the US used about 97.3 Quads of Primary Energy in 2016 [65], and after this primary energy has been transmitted and distributed and put to work you end up with effectively using 30.8 Quads and rejecting (wasting) 66.4 Quads of energy. This means that only 31% of the energy put into the system has been used efficiently.

Let's try to set a golden standard. We want to emulate countries with high social progress[66], and similarly a social progress when linked with energy availability per capita[67]. It didn't come as a surprise that Western Europe, Northern America, parts of Middle and South America, Australia, Japan and South Korea scored high on those indices. There are some minor discrepancies, especially when considering the gulf states, which have an unusually high energy per capita availability, but score relatively low on the social progress ladder. But we're looking for a gold standard. I've run my thumb down the list and arbitrarily considered Great Britain, the Netherlands, Denmark, Japan, and France. The lowest among them have a per capita energy consumption of ~4000 Watts per person ~ equal to 35,000 KWh of total primary energy per person per year. This seems like an awful lot, and

57

that's because everything is factored in. Not just the energy use in your home. But also, the energy used for your car, to manufacture it, to create the infrastructure to support it, the energy required for your hospital, police station, fire station, city council, road maintenance, buildings, services, goods, food, soap, water, etc. etc. etc. Suppose we would be able to distribute this kind of wealth fairly across mankind, we would end up requiring 385,000 TWh of primary energy for the entire world which is based on the consumption of fossil fuels at relative low efficiencies. Today we exploit about 170,000 TWh of primary energy. Remember, only 31% of this energy is used. By this measure we would use about 50,000 TWh effectively. Once we combust fuel, more than 60% of all the useful energy is expelled as heat through your exhaust thanks to the thermodynamic inefficiency of the conversion of heat energy to mechanical energy[68]. Unfortunately, we will never achieve a 100% efficiency in putting energy to good use. Energy losses beyond thermal are everywhere[69], even with technologies that do not rely on combustion[70].

Perhaps we could double the efficiency by eliminating the use of fossil fuels and replacing as many internal combustion engines and combustion generators with electric engines and generators. In some cases, this might not even be possible. Think about planes for instance. Or consider the sheer volume of cars. We might not be capable of transforming them all. Suppose we could convert half (half a billion cars), we would still have to figure out a way to keep the other half billion cars running with an internal combustion engine with fuels that do not have the same harmful footprint as gasoline and diesel have. But this will require fuel synthesis, and that's inherently less efficient, which means that we would require more energy, rather than less[71]. Future energy conversion scenarios are not that clear-cut, I don't believe that it will be easy, or perhaps even possible to get rid of all the combustion engines before 2050. But it sure is worth to pursue this goal deliberately.

I've heard it said that all we need to do is to replace all coal-burning energy and we're there. But that's far too simplistic and naïve. However, it would be a good start. Simply consider the fact that burning fuels has annual death toll of 7 Million People[72]. Beating that paradigm alone is worth the try. We should start somewhere, so let's start there. Simultaneously we can start working on replacing internal combustion engines when and where possible. Also, the use of natural gas as a heating fuel could be offset by using heat pumps, solar water heating, electrical heating elements, and centralized heat sources like highly efficient / high temperature reactors. Our entire heating infrastructure can be solarized, electrified or centralized. When talking about nuclear energy, using the energy that would normally be expelled would even be a better source for heat, dramatically increasing cost-effectiveness. This is something we're going to explore later in this book. Suffice to say, electrification and the use of waste heat from nuclear and perhaps geothermal have a significant role to play in the future if we want to beat our addictive reliance on burning and combusting fuels.

We are now going to hypothesize a cleaner energy future to show what effects some choices will have. Suppose we create a technological future in which we have doubled our end-use energy efficiency i.e. went from 30% to 60%, and only 40% is wasted during conversion. We also assume a total primary energy requirement of roughly 250,000 TWh (which is what you get if you extrapolate the figures from the 2017 International Energy Outlook[73]. Now we have all we need to start doing some simple calculations.

First, if we would consume 250,000 TWh at ~30% efficiency, we'd be using 83,000 TWh effectively. However, if we could double the efficiency, that would mean that we would need approximately 140,000 TWh of primary energy input (thus gaining 90,000 TWh in terms of efficiency). Let me remind you that this is a hypothetical situation. Also, note that this is a far cry from the world in which all of us benefit, and can live lives based on evenly distributed wealth in terms of energy availability, water, food, services, etc. For that additional tens of thousands of TWh are required.

The aim of the next exercise is simple. To generate enough energy to satiate an end-use demand of 83,000 TWh at an efficiency of 60% and thus requiring us to put in 140,000 TWh. How many generators would we need? First, we consider the correct metrics.

- Power/Capacity: is expressed in Watts (in this book with prefixes Kilo, Mega, Giga, Tera);

- Time: is expressed in hours / mostly the numbers of hours in a year i.e. 8766 hours.
- Energy: is expressed in Watthours (in this book with prefixes Kilo, Mega, Giga, Tera);

It therefore follows that if 140,000 TWh are consumed in a year, we require a capacity of 16 Terawatts. But that shortcut doesn't work here. No single energy source is available all the time, though some, like nuclear and geothermal, are available almost all the time. To account for the downtime, we add a metric called Capacity Factor (CF). People always argue with me about how to define CF. To me, it translates into this: CF is the amount of time in percentage that a specific generator delivers 100% of its maximum rated power. For instance, if you have a power source that is fully available 3 hours out of 10, it has a CF of 30% (or 0.3). It isn't that clear-cut however. We're talking about an average—measured over a year. There are moments when a power source operates at a fraction of its rated capacity. We consider the aggregate over a standard interval of a year, and then create an average to give a depiction of how well it performed. Capacity represents investment, CF represents investment return. A low CF equals stranded assets.

I've made the figure on the next page to show you the correlation between capacity factor (on the vertical axis) and capacity (on the horizontal axis) when you need to generate 140,000 TWh. The smaller the bar, the more available your power source is. You want to have a CF that is as high as possible, as that will enable you to deploy less capacity, which in term means less materials, which ultimately means that it is better for the environment.

Capacity Factor requirements for 140,000 TWh

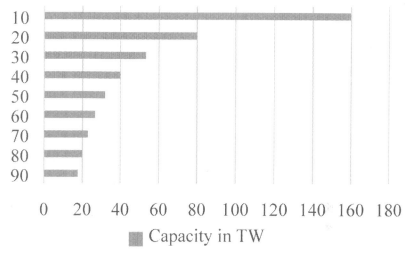

■ Capacity in TW

Now we can translate this into real-world figures. We are going to consider three technologies: nuclear (CF 90%), wind (CF ~35%), and solar (CF ~25%). And do note that these are the EIA figures[74] rounded up (wind and solar) and rounded down (nuclear). Also, these figures are quite optimistic, because if we consider Fraunhofer data for Germany, for instance, we may take note of far lower Capacity Factors. Solar CF ~10% and wind CF ~ 20%. We're now going to consider the material footprint of these technologies, first from a MWh per Ton perspective. This metric will show us how effective materials will be used. In the next figure we will see the most material efficient windmill (the Vestas V90 3MW[75]) - the best commercially available PV panel (SunPower X22 345 Watt[76]) - and an age-old generation II nuclear power station. The interesting thing to take away here is that the higher the bar, the more efficient your energy source is in terms of material input—bigger is better in this regard.

MWh/ton

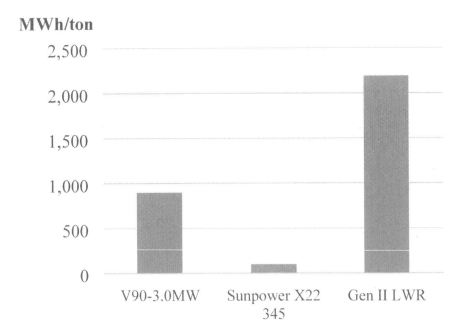

Remember that we wanted to build 140,000 TWh worth of annual energy generation? To maintain consistency, we're going to normalize to Megawatthours. 140,000 TWh equals 140 Billion Megawatthours. This gives us the following material requirements if we would create an exclusive energy generation infrastructure of each of these technologies

- Wind: 156 Million Tons
- Solar: 1.4 Billion Tons
- Nuclear: 64 Million tons

Even if we would add all the uranium, thorium, spent nuclear fuel, bomb-grade material, and depleted uranium, we wouldn't reach more than 80 Million tons. Slightly more than half of what is required to build a volume of wind generators capable of generating the same amount of energy in a year. I tell you this, to dispel the myth that mining for nuclear fuel would be an enormous burden. Even if we could and had to extract all of it, it would be dwarfed by the material requirements for any other energy technology. Additionally, it is important to point out that I omitted extra material requirements for solar and wind energy—think about transformers, transmission, etc.

And there are other considerations. The transmission and distribution infrastructure required[77] for wind and solar is much more elaborate. It requires a great deal of storage, additional power lines, and more complex grid distribution hardware[78]. That is not to say that we shouldn't build these technologies. On the contrary, we need every low-carbon technology to play a role in solving this problem.

If we would distribute more wealth over the planet, which is more than just money alone (but everything from water and food, to services and goods) we would end up more than doubling our energy consumption. We have a golden opportunity to expand on our energy capabilities without having to harm the planet and its biosphere. But it will require us to move ahead deliberately and thoughtfully.

In any case, providing prosperity and stability for all people on the planet is going to require far more energy than we consume today. Even if we could convince *the West* (OECD, etc.) to consume less energy, say a third of what we use today" and we help the developing countries to catch up, energy demand and the demand for raw-materials is going to rise regardless.

Wind & Solar

As far as short-term job creation is concerned, there's no technology more potent than Wind and Solar. The energy these contraptions capture is so dilute that we require millions upon millions of units to satiate any meaningful portion of our current electricity demand. And therefore, it logically follows that a lot of people are going to be needed to build this infrastructure and transport- and distribution networks. And this is often touted as the most economical model that is socially acceptable. But this begs the question: Is it the best model?

This might seem tangential, but I regard the mechanization of many menial jobs as a blessing for mankind. It opens up possibilities for people to engage in far more enjoyable and meaningful occupations. If we go down the path of mass-implementation of solar and wind energy, we will achieve a certain degree of regression. This kind of industry is not modern.

One of the glaring problems of wind and solar, the intermittency. These energy sources operate haphazardly. Under average circumstances, wind has a capacity factor of roughly 30%. And this means that it delivers its rated output for just 30% of the time, while delivering no output for 70% of the time. This means that for 70% of the time, the electricity required for a community must come from a different source. Worse, wind consumes power when not delivering it.

First thing to note here, is that a battery is not an energy-source. It's energy storage. So, before you can discharge any electricity from a battery, it has to be generated somewhere else first, then stored into a battery somewhere else.

Suppose we have a small city that requires 1 Gigawatt of power. Over a day, that would be 24 Gigawatthours of energy. These 24 Gigawatthours must be generated in 1/3rd of the day, which means that you have 8 hours to generate

24 Gigawatthours. How much wind capacity would you need? 24 Gigawatthours divided by 8 hours is 3 Gigawatts. So, you would need three times as much power capacity as you would actually require. Also, you have to store 16 Gigawatthours of electricity in order to fill the rest of the day. This is of course an optimum theoretical situation.

Realistically, we offset the downtime of wind generators with natural gas[79]. There's two kinds of gas generators[80] to note here. The Natural Gas Combustion Turbine (NGCT) with an efficiency of ~30% and the Combined Cycle Gas Turbine (CCGT) with an efficiency of ~60%. One would be inclined to couple a windmill to a CCGT plant, due to its higher efficiency. However, the ramp-up time of the CCGT is prohibitive. The NGCT can be ramped up much quicker and is therefore better suited to respond to the fickle nature of wind energy. In real situations, an NGCT never goes fully off, since it must be ready to make up for unpredictable wind and solar dropouts.

Dr. Robert Hargraves gave a presentation[81] at the Thorium Energy Conference in 2012 and I'm going to use his example. Suppose we have a factory that runs for 100% of the time and requires 1 Gigawatt of power. In 24 hours it would use 24 Gigawatthours of energy. A 1-Gigawatt wind facility would optimally produce 8 Gigawatthours in a day. This means that we would need 16 Gigawatthours from gas. Suppose we can run a NGCT facility for 70% of the time, so we would require an additional 1-Gigawatt NGCT facility to pick up the slack. Since the NGCT facility is only 30% efficient, we would be burning 55 Gigawatthours worth of natural gas to get 16 Gigawatthours of electricity. This equals 11 Million cubic meters of natural gas (1 cubic meter = 10.83 KWh)[82]. At 0.2 Kilograms of CO_2 per Kilowatthour[83] we would be emitting an additional 11,000 tons of CO_2. And, the NGCT burns about 30% of its full-power gas while waiting to fill in for waning wind/solar. This is called "backdown mode".

Suppose you would run a CCGT facility to create 24 Gigawatthours, The CCGT has an efficiency of 60%, and would require the equivalent of 40 Gigawatthours of natural gas to run all day. Coming in at 8000 tons of CO_2. Using a simple gas turbine to back up wind uses 55 Gigawatthours' worth of natural gas, while using a CCGT to provide all the energy uses 40 Gigawatthours of natural gas. In fact, the CCGT facility is better for the environment than the Wind and NGCT combination, as it emits less CO_2 per Kilowatthour.

We uncover a fundamental shortcoming of wind energy. We cannot run 100% of our electrical activities using wind energy, unless we overbuilt by at least a factor of three and add a significant volume of batteries or other storage technologies. Additionally, it is often claimed that through the distribution of wind generators over a large geological area ensures that "the wind will be available somewhere, at any time." But this would require a far greater installed capacity of wind generators than would be required in the first place. If we keep doing what we've done so far, coupling wind energy with gas energy, our greenhouse gas emissions will be higher than if we would use natural gas exclusively. It gets even worse when we consider Germany and take note of the fact that the ramping up and down of black coal energy generation coincides with wind and solar intermittency patterns[84]. So far, there's no reason to celebrate the implementation of renewable energy sources.

The exact same counts for Solar PV and Solar concentrated power plants as well. These technologies have capacity factors ranging from 10 to 25%. And do note that Germany has yet to hit double digits on their annual solar capacity factor (Data obtained from Fraunhofer directly—no external link). This has caused Germany to fail to meet its emission commitments to the rest of the world. So, offsets and backups are required. We're going to examine this later on. Materials are the most essential consideration when talking about energy technologies.

Before we continue, we have to address the issue of legacy. Solar panels and Wind generators are manufactured using many different materials and chemicals. neodymium, for instance, is an invaluable constituent for permanent magnets[85] and is used in large quantities in wind generators. The composition of a permanent magnet is roughly 68% iron, 29% neodymium, 2% dysprosium and 1% boron. A paper[86] published in MDPI called *"Material Flows Resulting from Large Scale Deployment of Wind Energy"* provides some proverbial straws in the wind. With an estimate of 27 Gigawatts of installed wind capacity in Germany by 2050, they would require 2,300 Tons of permanent magnets, or 667 tons of neodymium. This gives us an average of 24 tons of neodymium per Gigawatt of capacity. This seems low, but we have to consider the fact that there are also viable windmill designs that do not require permanent magnets. If we would implement Jacobson's energy mix, this would amount to 13,000 Gigawatts of wind generators spread across the planet, containing a conservative 312,000 tons of neodymium.

Why does this matter? The damages from mining and extracting neodymium from monazite[87] and bastnaesite ores are immense. Most of our rare earths come from China, it's not so much the fact that we have to extract the rare earth metals, but the fact that the Chinese have so little environmental regulation, that the byproducts of mining and processing these elements are simply thrown away into a big tailings-lake, causing all sorts of problems[88] for the local environment. And there's no way of competing with this industry, because they don't have to pay the environmental costs, they just let the environmental externalities run their course in a big lake. Also, their labor is much cheaper[89], and thanks to their monopoly they can price these commodities whichever way they want. I would be more in favor of rare earth element extraction if the Chinese monopoly would end and Monazite extraction and processing would be done in countries with better environmental standards and higher efficiencies.

The interesting part about Monazite is that it contains a plethora of valuable elements. Apart from the rare earth elements it also typically contains 6 to 12% of thorium. At this moment, mining and processing of Rare Earth's (R-E) from Monazite in countries outside China is seen as a liability, because it contains so-called radioactive / nuclear "source material". Here's the definition of source material as provided by the US Nuclear Regulatory Commission[90].

"In general terms, "source material" means material containing either the element thorium or the element uranium; provided that the uranium is not enriched in the isotope uranium-235 above that found in nature. Both natural uranium and depleted uranium are considered source material. Source material can also be a combination of thorium, depleted uranium, and natural uranium and the material can be in any physical or chemical form. Ores that contain uranium, thorium, or any combination thereof, at

one-twentieth of one percent (0.05 percent) or more by weight are source material."

This will put the miner/refiner in legal jeopardy because of thorium tailings and waste management. A way to solve this problem of how to hand source material would be to create a thorium bank[91]. This idea is spearheaded by John Kutch and Jim Kennedy of the Thorium Energy Alliance. They basically foresaw a repository for thorium, keeping it in a federally regulated building for those who contract to use it. This thorium bank would be a means to open R-E extraction and processing for countries outside China. Not only are these elements required for the permanent magnets in wind generators, we also use them in a multitude of different technologies such as power tools, optics, display screens, electric vehicles, catalysts, speakers, batteries, smartphones, etc. In effect, these elements are essential.

If we want to minimize damaging effects from mining/refining, we must end the monopoly of China. We need to expand R-E activities to countries where governments and mining/refining companies maintain a commitment to excellence. This would minimize the negative effects from laissez-fare mining as exists in China. The problem is that this seems incongruent with certain economic principles. By its very nature, more regulation, means more costs, and therefore the cheaper competitor wins. However, I've shown you that there are measures that can be taken to undermine these monopolies. It's also important that this neodymium issue is one of many. All extractive activities should be subject to high quality standards. Not just of product, but also in terms of how we leave the land after we've extracted the valuables.

Let's consider arguments in favor of solar and wind. In general, solar operates optimally at places where it can become quite hot and solar irradiance is sufficient. There, it can be relatively reliable for partially offsetting day-time energy requirements to power domestic, office and commercial/industrial activities. That often includes a large load for space cooling in these hotter environments. When panels are placed on existing residential, commercial and industrial roofs, the power source is close to the demand and this can bring benefits in managing the electricity network (as well as some challenges). Another particularly beneficial application in sunny climates is collecting solar energy from rooftops to direct heat water for residential and smaller scale commercial uses. In these hot and sunny places, electricity demand often peaks in the early evening. Solar panels facing the direction of the setting sun (east in the northern hemisphere and west in the southern hemisphere) can lower peak demand, which eases the strain on the transmission and distribution network and will potentially displace fast-ramping and polluting natural gas turbines. This, and solar hot water heating, is a really sensible way to use the sun's energy to help meet the climate change challenge and are completely complementary to the use of nuclear power in making a stable, reliable and clean electricity grid. There are good arguments for using roof space for PV and/or solar heating systems in sunny climates. It's precisely within this context that they make the most sense. However, building grid-scale power plants, comprised of PV technology, is less-efficient use of precious space and materials to do a job that would be better done using nuclear fission.

Concerning wind, I am not arguing that wind technology is bad. In many parts of the world, the early stages of a wind generating sector has helped to reduce greenhouse gas emissions. I am merely identifying that the variability of wind becomes a genuine and significant challenge especially if we try to use too much of it compared with other clean energy options. Like every clean energy technology, it is going to have a place in this massive challenge. Our job as concerned environmentalists is to see technologies used according to their advantages, and not pushed into places where we start making the more important job artificially hard for ourselves'

Coal, gas, oil, and biomass require smokestacks to relieve their furnaces from CO_2 and other agents that would hamper the efficiency of the burn. Note that no energy source will be free from carbon emissions if we need fossil fuels to mine and purify the materials, manufacture the technology, transport it, install it, maintain it, and decommission it. And this is where we can start to look at technologies that can help us put the use of fossil fuels into decline. It starts with replacing the big emitters at the top. And moving on from there.

Anything that has a furnace is out, which means that we are left with nuclear, geothermal, wind, solar, and hydro technologies. Wind and solar have a role to play. Solar as a grid-augment for daytime activities, and wind as a source for energy in more remote regions on the planet. We must accept that there are limitations. In a grid-context, baseload capabilities cannot be ignored. And when punching out coal, gas, and oil, nothing beats high capacity factor energy sources like nuclear, hydro and geothermal. All of these can be made to load-follow. But it wouldn't make any sense to prioritize wind and solar energy over nuclear, or geothermal. Wind and solar are additions, not cornerstones, because they are inefficient uses of materials and the environment. Their energy is anything but *"free"*.

Part three: We can solve this problem

Eliminating misconceptions

Dispelling actual myths is tiring work. Especially when people are so invested in them and have become cognitively closed to reality. However, I think that it is important to bring you up to speed regarding these arguments. Because the truth is far more nuanced. We're going to consider popular misconceptions about nuclear energy.

1. We will run out of fuel for nuclear reactors;
2. We cannot build storage facilities that last long enough
3. Building nuclear reactors increases proliferation;
4. Nuclear energy is dangerous;
5. Nuclear is slow and expensive.

Myth: We will run out of fuel for nuclear reactors.

As it stands, roughly 440 reactors are in operation today, most of which are of the LWR type. The fuel for these reactors is so-called low-enriched uranium (LEU). Natural uranium consists of 99+% uranium238 (U238) and about 0.7% uranium235 (U235). In a thermal reactor you need to increase the amount of U235 up to somewhere between 3 and 20%, but typically around 5%.

And this brings about one of the issues that is often raised when talking about nuclear energy. People think that because we predominantly use U235 to fission in our LWRs, we're going to run out. But this depends on many different factors. The first is the assumption that nothing is going to change in the world of nuclear technologies; That all we will be using are LWRs and a couple of other designs. In fact, back in 1962, President John F. Kennedy already knew this to be false thanks to a report written by Glenn Seaborg.

Where it is true that uranium is a finite resource, it is also true that we have many options to use the more ubiquitous U238 and perhaps Th232. Once we tap into those resources we no longer need to worry about depleting uranium or thorium. The fuel consumption efficiencies will go up dramatically and we will have a tough time depleting the resources that have already been mined. Let me show you the numbers.

According to the *2016 Nuclear Energy Agency, Uranium Resources, Production, and demand report*[92] we can recover about 7.6 Million tons of uranium on our planet. These are conventional resources. And we've already unearthed roughly 2.7 million tons of uranium. If we consider these figures it surely looks like we're going to run out. But here's the essential thing to take away. Of this 2.7 million tons of uranium only 3 to 5% has been used, which means that we already have a huge pile of uranium.

Our current rate of uranium Production from extraction is approximately 55,000 tons per year. This means that the known uranium reserves would last for another ~140 years, if we keep continuing to use uranium as we do today, without any innovations or changes in the rate of use. If we add more LWRs, we decrease the amount of time before our known uranium reserves are depleted. Do note that there are better explanations out there, I'm taking this trough from a sum-total/non-cost perspective. Consider for instance Ripudaman Malhotra's Cubic Mile of Oil, which is an entire book dedicated to the subject of resource consumption. Also, it is important to note that not all recoverable uranium has been discovered, and that the oceans contain another 4.5 billion tons of uranium[93]. Which means that it is in virtually limitless supply (more than 81,000 years). However, means to extract it from ocean water are still in development. Consider for instance Masao Tamada's 2009 paper on uranium extraction from seawater[94]. Do note that it will take more energy to extract this resource, it is preferred to find resources elsewhere before we turn to the seas for our fuel.

Additionally, there's the possibility to breed=fuel, as JFK knew. This means that we no longer depend on U235 but can also use U238 or Th232 as a basis to create fissile fuel. This can be done both in the (moderated) thermal-neutron spectrum, with thorium, and in the (unmoderated) fast-neutron spectrum with thorium and uranium. In any case, a ton of Th or U can power about 1 million homes for a year—$E = mc^2$ and Avogadro's number at work! No other energy source comes close, except light-element fusion.

First, let us consider the amount of thorium at our disposal. The World Nuclear Association estimates that there is at least 6.5 Million tons of thorium[95] available in recoverable deposits. India has some particularly rich deposits of thorium and is creating reactors that use thorium as a fuel. So is China—thorium is a waste product of their massive rare-earth mining and refining industry. Fuel utilization of thorium or uranium in breeder reactors is close to 100%.

To figure out how much energy we could get from these uranium and thorium resources if we could use them for a full 100% we have to do some calculations. Typically, a single atom's fission yields about 200 MeV (Million electron Volts)—a tiny amount of energy. But remember that we are splitting everything, and I've taken the molar mass of U238, this means that there are about 4 moles of uranium in a kilo. After some wizardry with incredibly large numbers (including Avogadro's) we discover that 1 ton of U238 (if all of it fissions) yields about 21 TWh of thermal energy.

We have roughly 7.6 million tons of uranium, so if we multiply that by 19 TWh per ton we get 167,000,000 TWh currently remaining in the ground. Additionally, we extracted 2.7 million tons. If we add those, we would get 223,000,000 TWh. Suppose that we would require 300,000 TWh per year, these resources would last for 750 years. Double that to 1500 years if we start using the 6 million tons of thorium as well. And that's not considering the sheer volume of uranium that is present in the oceans which will last for hundreds of thousands of years, and that's discounting the fact that it is being replenished continuously through tectonic activity.

We must also consider that we're probably not going to rely on nuclear energy exclusively, but will also implement a significant volume of Geothermal, Hydro, Wind and Solar. People are going to press me for fusion, and I think that it has potential. But I'm very tentative and cautious. My caution stems from analyzing the availability and ability to extract certain valuable elements needed for all these technologies. The hubris surrounding clean energy sources clouds judgement. Where I think that fusion is feasible, it still has not been proven to be available commercially, and we don't yet know the material and feedstock footprint. Not until we know those metrics can we honestly assess their merits. This is different for Generation IV fission reactors which are all based on proven principles.

We may therefore conclude that we have a nigh unlimited source of clean energy. Anything past a millennium from now may be considered unlimited, because before that time has passed, we either have gone extinct thanks to our inability to combat ocean acidification, pollution and global warming effectively, or we will have started extracting valuable materials from other planets and moons and planetoids—the Moon and Mars have lots of thorium.

Let's consider this: with breeding, we could use 100% of uranium fuel, whereas today we use about 5%. If just that were true, we would be able to increase fuel efficiency 20-fold. If that's not worth pursuing, nothing is! With thorium breeding, we get even more. Furthermore, Gen-IV reactors run at higher temperatures, and thus yield about 50% more output energy from their turbines, etc. than do present water/steam reactors. Fission energy is abundant, lasting, and, quite natural[96].

Just in case people start doubting the feasibility of such an energy scheme, consider the fact that the principle that makes this possible already exists. It is called a breeder reactor. We know that we can do this. Also, the technology to stretch the fuel efficiency is in development.

The expedited development and deployment of breeder technology is fundamental if we want to achieve long-term deep decarbonization for a sustained future. This is what JFK was told in 1962. And that's what led to MSR development at ORNL in the 1960s.

Myth: We cannot build storage facilities that last long enough

What to do with nuclear waste? Particularly the hot stuff that comes out of reactors. At least that's how some people picture this conundrum. How we answer this question is predicated on whether we are smart enough to move forward on breeder technology and chemical reprocessing (of the fission products). If we start deploying more breeder reactors, storing nuclear waste becomes a total non-issue. Both in terms of time as in volume.

74

First, we're going to examine what the *waste* is composed of.

We started with ~95% U238 and ~5% U235—the U238 is mostly for dilution. Out of 100% of the fuel we put into the reactor, roughly 5~6% changes into something else during the five years it spends in the reactor. Some atoms split, and leave fission products, some atoms absorb neutrons and transmute into something heavier (e.g. U238 moves to plutonium and heavier), and some of those transuranics split as well, leaving similar fission products. All the splits give us clean thermal energy.

The reason why we cannot keep using the fuel is because of the buildup of fission products that absorb neutrons, like Xenon. This diversion ruins the neutron *economy* in the reactor and means that the number of neutrons required for a chain reaction would drop below a critical threshold, shutting fission off if you would leave such a solid fuel in the reactor. Such fuel is now considered *spent*. This is one reason why molten salt was developed at ORNL to carry fission fuel—gasses like Xenon simply bubble out to allow capture. Other neutron absorbers like Hafnium can simply be chemically extracted from the salt as needed.

1. Uranium and transuranics
2. Strontium and cesium
3. Technetium and iodine
4. Other fission products

Fission products are very radioactive because their nuclei have more neutrons than they wish, so they decay quickly by emitting electrons (beta particles) and gamma rays (stronger than X-rays). Some decay in seconds, others in days, others in years. The reason we store spent fuel is to allow these isotopes to decay to normal elements and not get into the outside world any sooner. Cesium, Iodine, Strontium are dangerous ones because they appear to living things as 'food'—Cesium mimics Calcium and is taken up by bone cells, etc. The other fission products are interesting but pose no concern for storage. Technetium has a long-lived, relatively harmless, isotope. Technetium will be radioactive for about a couple of hundred thousand years. Strontium and Cesium decay by half in about 30 years, so we want to store them safely for 100+ years. And the rest is either fissile fuel, or fertile isotopes to become fuel. These elements are not waste, but fuel energy for other reactors, and are comparatively harmless—solid in form.

What to do with this stuff that has been mislabeled as waste? We can throw it away, which many want to do. Throwing away means hiding it in a cave, where it will linger forever. The argument against it is that mankind will not be around to take care of it for long enough, and some distant civilization might stumble upon these caches of nuclear waste and unleash it's deadly spectre on whoever lives on the planet at that time. H.G. Wells couldn't have written it any better. Granted, it is likely that mankind won't be around anymore by that time. But that still doesn't mean we cannot dispose of it safely. However, we need to things differently. What other options do we have?

Let's consider what happens to solid fuel first.

We're talking about solid fuel pellets. This means that all the fission products are either trapped in the pellets themselves (inside the fuel cladding). The fuel cladding is a closed Zirconium tube surrounding the fuel pellets. And all these tubes, filled with fuel pellets, are held together by a metal fuel assembly. A reactor core contains many fuel assemblies at the same time. Every 18 months or so, these assemblies are shuffled around so their fissile atoms get evenly exposed to neutrons flitting around the core. Those in the middle experience higher neutron flux than the ones on the outside.

Once the fuel is considered spent, the fuel assembly is lifted out of the reactor, and submerged in a water-filled spent fuel pool, in the reactor building. The fuel pool is there to cool the fuel pellets while the initial decay heat from fission products dissipates. After a few years have passed, the spent nuclear fuel can be taken out of the fuel pool and put into an airtight dry cask storage container. This is a temporary solution, and that's good, because we want to be able to recover what is inside—the fertiles & fissiles.

As mentioned earlier however, some countries have different philosophies on how to go forward. Let's consider Finland. They will probably be the first to complete a permanent waste storage facility, and first to start using it. This facility is situated on the island Olkiluoto, just off the Finnish coast, which it shares with three nuclear power reactors, one of which is an EPR that is still under construction. The storage facility is built deep within the island's granite bedrock and is supposed to be permanently sealed by 2120[97].

There's also a consideration to be made about the way the Finish are planning to store this waste. The KBS-3 containers[98], that will be used to store the spent nuclear fuel are made out of thick copper—chemically similar to the surrounding rock content. Once these have been stored underground, and the facility has been closed, it would be a real waste of precious fuel and other isotopes—also wasting a prime resource in the electrical power world.

I would be fine with these kinds of waste facilities, if we would store elements in there that needed to be stored. In truth, some of the fission products deserve to be sequestered. Think about the previously mentioned Strontium, Cesium and Technetium, unless we can find uses for them. Remember, an isotope is only waste if we waste it.

Most other fission products are either completely harmless or could be used if we were able to extract them quickly enough, depending on how stable they are. More than 90% of the spent nuclear fuel is still useful, as fertile or fissile, and should not be thrown away.

I feel confident to claim that our current means of storing spent nuclear fuel are sufficient, and should not be expanded, unless it is warranted in terms of volume. We should not be wasting resources which we have worked hard for to obtain and purify. Suppose we do want to sequester spent nuclear fuel, I think that we can build a facility that lasts for millions of years. Boring a deep hole in the ground and plugging it is not a high-tech problem.

Myth: Nuclear means bombs

The truth is far more nuanced than odd statements by those who propagate anti-nuclear dogma. One of the best resources in this regard is an article[99] called *"Why nuclear programs rarely lead to nuclear proliferation"* which was written by Nicholas S. Miller and published in November 2017, in the International Security Journal.

Here's what little technical background I can give you. Try to find more information and learn about these issues so that you can refute any *"but the bombs"* argument that you could encounter in the future. To build nuclear bombs we need a fissile: U233, U235, or Pu239. Weapons-grade U233 and Pu239 (respectively from thorium & U238) can be bred in special reactors, whereas U235 can be extracted from natural uranium using specialized high-end centrifuges[100] (as used by Iran).

It is important to point out that U233 or Pu239 that could be extracted from reactors after being used in a normal fueling cycle, are contaminated with U232 or Pu240. U232 has a decay product which is a strong gamma emitter that makes the U233 too hot to handle, while sending out a clear signature which makes it easy to track. Pu240 on the other hand is a prohibitive contaminant for Pu239 bomb builders—it fissions unpredictably. The IAEA specifies how much of these contaminants in a fissile load makes it proliferation safe.

Magnox[101] and RBMK[102] were two reactor designs specifically engineered to create plutonium to be used in nuclear weapons. Additionally, these reactors generated energy for civilian purposes. And that's one of the reasons why nuclear weapons and nuclear energy got intertwined. These were military reactors that were operated on a completely different refueling regime, only leaving the fuel intended to be *toasted* in the reactor for a maximum of three months[103]. If one left the fuel in there longer—it always is—the buildup of Pu240 would inhibit the construction of a nuclear bomb. If the amount of Pu240 would be too high, it could cause premature fission reactions and render the bomb useless or far less potent—*a fizzle*. So, the refueling regime is quite strict if you want to breed weapons-grade plutonium from natural uranium. That's what the RBMK reactors at Chernobyl were designed for. Ordinarily, what comes out of a Uranium reactor is called reactor-grade plutonium and is not suited for weaponization, but this Pu239 will generate clean fission power when reprocessed into so-called Mixed Oxide fuel (or MOX). For Th-U233 reactors, their U232 content can be spiked in different ways, such as including some Th230, which will breed U232.

Under the 20-year Megatons to Megawatts program[104] The US and Russia converted roughly 20,000 warheads worth of weapons-grade uranium into

15,000 tons of low-enriched uranium fuel used to create clean energy for the US Economy. Conversely, the Russians have had experience running the BN600 and BN800 fast-neutron reactors[105], which are capable of breeding Pu239 as fuel but have also been used for *burning* weapons-grade plutonium. nuclear reactors are the most effective means to get rid of our vast stockpile of nuclear weapons. And if there's any hope of getting rid of these, we better embrace civilian nuclear energy and keep pushing for more "Megatons to Megawatts" multi-lateral agreements, thus, creating olive-branch moments between nations who have been at odds for decades and providing some peace of mind to mankind.

Let us acknowledge that the application of the science of the nucleus, a part of the physics discipline, is not inherently evil or good. It is up to us to use it for good purposes. It has become clear that the nuclear bomb is perhaps one of the most evil contraptions we've ever devised. Fortunately, we've managed to minimize its use against civilians. The markers at Hiroshima, Nagasaki, and all sites of war serve as a stark reminder that we must work out our differences diplomatically rather than through strength of arms. To minimize the disparity in the world, we need more clean energy, and clean nuclear energy has been proven to be a force for good, that can eliminate the threat of nuclear weapons, while simultaneously powering high-tech, industrious and well-faring nations.

Myth: Nuclear is dangerous

This follows in the wake of the fear of nuclear weapons proliferation. We're talking about fear of radiation, for instance. Entire books have been written about this subject, and I'm not going to do a complete in-depth analysis of the problem because there simply isn't enough space in this book to begin. So, I'm going to give you my view, which I derived from years of reading up on WHO and UN documents. Here's the lowdown.

First, radioactive elements and ionizing radiation are ubiquitous, and our bodies cope with them daily from birth, as does all life now, and as all life has for all its existence. Think about the continuous exposure to cosmic rays; eating fruits and vegetables that contain radioactive potassium40 and carbon14; swimming in the ocean (containing billions of tons of Uranium ions); preparing sandwiches on a radioactive granite kitchen top; Flying at cruising altitudes; Getting an X-ray; The list goes on. Radiation and radioactive elements are everywhere. We are subjected to them with or without nuclear energy because they simply exist in nature. And, there has always been stronger radiation in the past, for the simple reason that radioactivity arises from radioactive isotopes decaying to stable elements, as "half-life" means & measures. U238's half-life is about 4.7 billion years, so today there's about half as much U238 as there was when the solar system was forming. Thus, U238's contribution to our radiation exposure today is far less than early life here experienced and obviously dealt with.

If there's an accidental release of radioactive elements from a nuclear power plant, as we've witnessed at Three Mile Island, Chernobyl, and Fukushima, it is far less damaging than has been presumed. In fact, Three Mile Island and Fukushima had negligible health effects. Chernobyl has indeed had health effects, partly due to Soviet era secrecy and delayed help to citizens to combat Iodine131 promptly, but the scope remains limited, and the area around the plant has rebounded. In fact, some residents never left, and the exclusion zone has become a wildlife refuge, sustaining large animals like the European bison, wild horses, wolves, boar, deer, etc. It is also becoming an area that attracts thousands of tourists each year. People interested in the post-apocalyptic nature of the area, those who are nuclear energy aficionados, biologists, engineers, and so on. I want to visit Chernobyl in 2018 or 2019.

Second, the hypothesis that there is no safe dose of radiation is incorrect. Life itself confirms that. I'd like to reference the books *"Nuclear is for Life"* and *"Radiation and Reason"* by Wade Allison (Emeritus Professor of Physics at Oxford) for more information on this subject. The evidence in favor of the so-called Linear No-Safe Threshold hypothesis is very weak. Some learned people suggest that low doses of radiation are essential, or have beneficial effects, like immunizations. However, solid evidence is yet to be determined. I suspect that there's a gray area in which the possible beneficial and malignant effects of radiation overlap and no real demarcation point can be established. That said, too much radiation at once, can be very harmful, because it can disrupt key molecules in our cells. This, of course, is why targeted radiation is an anti-tumor therapy. There are areas in nature where background levels of radiation are much higher than those inside nuclear power plants. Think about the famous tourist attractions in the Lodève region in France, or Guarapari Beach in Brazil, or Denver. But also consider the fact that astronauts are willingly subjected to extremely high doses of cosmic radiation, and as far as I can tell, no increased morbidity from cancers can be established in these kinds of extreme occupations. In any case, there's a lot of debate. I am not afraid of radiation. It is up to you to rationalize it. My advice, seek out evidence and compare different models. Don't listen to so-called experts who are out there spreading unjustified narratives of fear and damnation. The world is not turning into a radioactive wasteland. In fact, the opposite is true because it is getting less radioactive with every passing year due to the natural decay of radioactive isotopes.

Third, the radiation from nuclear power is contained within a robust containment vessel—robust primarily to contain vast steam pressure should the plumbing in a conventional LWR fail. Even if breached, it is highly unlikely that people will be hurt.

The Chernobyl accident in 1986 was a steam explosion within the reactor core itself, propelling fuel and fission products into the air above. Because the RBMK lacked true containment it was thus illegal everywhere outside the old USSR. Iodine131 especially went medically unaddressed and caused some thyroid malformities in the Ukraine. Some became actual cancers. Fortunately, thyroid cancer is easily treated. So, very few people have died from its effects. The main death toll at Chernobyl was borne by the courageous workers trying to stop fires in and around the reactor core's remains.

Fortunately, no such reactor has ever been built since and all other reactors are completely different in terms of safety.

Fukushima has yet to claim its first radiation victim. So far, all accounts of radioactive Pacific Ocean pollution from Fukushima have failed to appear. In fact, we should be far more worried about the excess man-made carbon dioxide and heat being absorbed into the ocean, as we know that these are causing major disruptions in the realm of phytoplankton, ocean food chains and coral reefs.

Some people liken Chernobyl and Fukushima to the worst environmental disasters ever to have happened on the face of the earth. But the facts are quite the opposite. If we consider the scope and real effects of these accidents, we may call them nothing more than expensive industrial mishaps with little to no real danger to the rest of the world. But amidst all the spectacular reporting and the "fear-porn" actors and movements it is hard to see the actual scope of these issues. Don't let your judgment be clouded by hoopla and hubris.

And that's where I want to leave you. Nuclear-power safety is a very interesting subject, with a rich history in practical events. That's not to say that it went wrong, but that there were moments where we have learned a great deal. It's much more fascinating to consider what good nuclear energy has brought humankind so far. Simply consider that Fukushima. Before it failed, had generated enough clean energy to save thousands of Japanese lives from other energy-source pollution. Three Mile Island caused reforms in US nuclear-plant operations that saved even more lives from pollution by increasing US nuclear output from existing plants by about 40%, eliminating the need for any equivalent coal power.

As a flourishing humanity and the availability of energy are directly correlated it is safe to say that the existence of roughly 440 operational nuclear power reactors has brought wellbeing to hundreds of millions of people for over five decades. Civilian nuclear energy has done so with low carbon emissions (mainly from secondary processes), while carrying local economies on its broad shoulders offering hundreds of thousands of people high-paying jobs and countless of spin-off benefits to local economies.

Myth: Nuclear is slow and expensive

People who claim that nuclear power stations take a long time to build, are cherry picking facts. When we consider the Olkiluoto, Flamanville, Hinkley C, Vogtle and VC Summer projects, they are right. And there are plenty of other projects that took between ten and twenty years to complete. So, if we consider this argument, we have to take a more nuanced perspective. A nuclear powerplant can be built and brought online quite quickly. I can cherry pick more than 67 nuclear reactors that have been built within 5 years. We may also note that 266 reactors have been completed in a timeframe between 5 and 10 years. What we learn here is that a power plant can be built rather quickly, but that there are prerequisites that need to be fulfilled before we can streamline the construction of these power plants. In the next chapter we will examine Michael Shellenberger's seven secrets to deploy contemporary reactors in a timely fashion. But now we're going to focus on the cost aspects, and the ancillary benefits that come in the wake of the deployment of civilian nuclear power plants.

The best paper to cite here is Peter A. Lang's *"Nuclear Power Learning and Deployment rates: Disruption and global benefits foregone."*[106]

Peter Lang writes: *"The fact that fast learning rates existed up to about 1970, and in South Korea since, suggests they could be achieved again..."*

...The US's post-reversal learning rate was the worst of the seven countries. The reversal occurred two to four years later in the other countries and the real cost increase was not as severe as in the US. This suggests the US may have negatively influenced the development of nuclear power in all seven countries (and probably all countries). It also shows that technology learning and transition rates can change quickly and disrupt progress, in this case delaying progress for about half a century so far...

...When cumulative global capacity of construction starts was 497 GW, the OCC [Overnight Capital Costs] of nuclear power would be 5% to 15% of what it was in 2015 (except in India); for example, the OCC would be $349/kW in the US, $257/kW in France, and $484/kW in Japan (Table 3). These are much lower than the OCC of fossil fuel and other alternative electricity generation technologies...

...If the pre-reversal learning rates and the Linear and Accelerating deployment rates had continued, the OCC would be approximately 2% to 10% of what it was in 2015 (except in India where it would be 31% to 33%) ...

...These are striking cost reductions that, to be achieved, would have required pre-reversal learning rates and deployment rates to continue. If

the rapid learning and deployment rates that prevailed pre-reversal could be achieved again, nuclear power would become much cheaper than fossil fuel technologies in the future. Some may regard this as too optimistic. However, there is no apparent physical or technical reason why they could not have continued and cannot prevail again. They have prevailed in South Korea over the past 40 years."

Also, if we consider the 2010 OECD Projected Costs of Generating Electricity report we may take note of a very wide range of different costs in nuclear construction, fuel cycle costs, decommissioning costs, operations and maintenance costs and so on. Costs[106] can come down as low as 30 USD/MWh in China for the AP1000 and CPR1000; Korean APR1400 and OPR1000 come in at 42 and 46 USD/MWh.

Lang suggests: *"The fact that rapid learning and deployment rates prevailed in the past suggests they could be achieved again. To achieve them, it is suggested four steps are needed:*

- *First, recognize that the disruption to the transition occurred and the impediments to progress continue to this day.*

- *Second, recognize the consequences of the disruption for the global economy, human wellbeing and the environment, and the ongoing delays to progress.*

- *Third, identify the root causes of the disruption and the solution options.*

- *Fourth, implement policies to remove impediments that are retarding the transition.*

The benefits forgone cannot be recovered, but future benefits can be increased by amending the policies that caused the cost increases and slowed the deployment of nuclear power. Human wellbeing could improve faster if the impediments that are slowing the development and deployment of nuclear power are removed."

This observation is most critical of all. This negative trend starts at the legislative level, and that is where we have to start reversing it. Lowering or removing the barriers to nuclear power development can lead to an increase in the fair distribution of wealth through increased high-tech / high-energy nuclear developments and innumerable spinoff benefits. It is likely that in the wake of developing new nuclear power generating capacity we would see a dramatic increase in overall health from cleaner air; We would see burgeoning communities thanks to the long-term stability that comes in the wake of the construction and completion of a nuclear power plant; overall, the benefits of these developments would overshadow those of any fossil fueled technology. It would be orders of magnitude better than coal and gas in terms of providing excellent / well-paying jobs and long-term stability. When we combine this with some hydro, geothermal, solar PV and solar heating & cooling and storage, we can create a versatile energy infrastructure that is better suited to cope with day to day variations. These claims will be further substantiated later in this book.

Contemporary nuclear energy

The light water reactor (LWR) is the dominant technological design with which we create energy from splitting atoms. If we take it through from a simplified perspective, the LWR is a big steel barrel filled with water and fuel assemblies, filled with fuel pellets consisting of U238 and U235, plus metallic rods to control the critical abundance of neutrons. Once the control rods are retracted from the reactor, neutrons can move freely through the reactor core. The water serves as a moderator, which means that it slows neutrons down, they will more easily fission fissile (U235) atoms. This means that the reactor operates in the thermal-neutron velocity spectrum. Because an atom's nucleus is held together by the strong nuclear force, once this force is broken, a part of nucleus' stored energy is released as heat into the surrounding medium. The water in the reactor serves as the medium (working fluid) to transport this energy to a heat exchanger, perhaps a steam turbine and then to a generator where it is turned into electricity.

LWRs are so robust and reliable that they can produce copious amounts of energy for decades. They have capacity factors ranging from 70 to more than 90%.

There are many slight differences among Light Water Reactors. But we have two major successful design principles: we have Boiling Water Reactors and Pressurized Water Reactors. It is also possible to use different working fluids and moderators. For instance, there are Graphite Moderated Water Reactors, there are Pressurized Heavy Water Reactors (Canadian CANDU), there are gas working fluid High-Temperature Reactors and finally we have the very interesting sodium (or even Lead) working fluid Fast Breeder Reactors. Russia has some very good experience with those, and is first to commercialize a design, the BN800.

We've learned that robust containment buildings are a requirement when building LWRs, because of water's limited temperature range (100C) between liquid and gas (steam). And, we don't take for granted things like earthquakes and tsunamis. Consider Japan, for instance, they are now slowly, but surely restarting their nuclear reactors after having re-evaluated their facilities via a new regulator, and they improved whatever needed to be improved.

It is fair to re-establish a robust sense of confidence in current technologies. The existing fleet of nuclear reactors is operating excellently. No energy technology is as carefully regulated as is nuclear power. Aside from that, it is essential to re-invigorate and restart nuclear developments and deployments. First, we need to keep our current fleet of nuclear reactors operating. We can't afford to lose precious carbon-free electricity generation. Second, we need to keep building contemporary nuclear reactors. But for that to happen, things must change. We've created an environment in which building a new reactor facility has become increasingly more difficult, except perhaps in China and a few other counties. We must change this paradigm to start taking out coal and gas fired energy generation.

Michael Shellenberger, Executive Director of Environmental Progress, wrote an excellent essay in which he postulates seven secrets to cheap and abundant nuclear energy. Let's see what we can learn by addressing each of these points individually.

One: Build National Consensus around a long-term energy plan.

Apart from the obvious, this is the first important point. Especially the issue about committing to a long-term strategy. Sentiments in Politics are fickle. Often long-term visions are trumped by the possibility of short-term gain. How to get into office, how to secure a prestigious well-paying job after your term, or less insidious motives. It might be the quest for some short-term victory over ideological opponents. Rarely do you see politicians try to build up a solid framework that will ensure stability for decades to come.

87

And yet that's precisely what nuclear offers. Stability for decades. If you commit to long-term nuclear development and deployment, you ensure job-safety, economic, educational & environmental benefits, plus power-grid reliability for decades and more.

And that's where Shellenberger's second point comes in.

Two: Engage the public

What does nuclear have to offer? Some might say fear and damnation. But this hardly what follows in the wake of the deployment of a nuclear plant. Fact of the matter is that one single nuclear power plant keeps thousands of people occupied for decades. Some people like to point out that the economics of a nuclear power plant are bad, whereas the exact opposite is true. Apart from being an energy generator, it also enables people by helping them get meaningful, good paying, stable jobs all around. From government jobs, to people working on the plant, to additional jobs required to keep the plant going, and keep the community thriving. When you build a nuclear powerplant, someone else will build a super market, and a gas station, and a pharmacy, and malls and shops and restaurants, and so forth.

It is often said that renewables are the biggest job generators. But when you consider NEI studies concerning the economic benefits of nuclear power stations, you can attribute a similar success to nuclear power, as we do to renewables. The only difference here is that the level of expertise required— a worker's education and productivity (energy produced per worker) in the nuclear sector is much higher, so even wages and the local economy are better.

Shellenberger states "Engagement efforts must be science based, informed by the best psychological, sociological and public opinion research available."

Ask yourself the question: "What will nuclear bring me and my family and friends, other than just energy."

Three: Standardize to a single design

We used to build one-offs. Which meant that each reactor needed to be designed and build according to varying specifications. This is not a smart way to go forward. You want to build each unit the same. With as much off-the-shelf components as possible. This is a way to ensure a steep learning curve in nuclear energy.

Once you achieve this, you can move forward and manufacture components, rather than build them. Your reactor should become a glorified IKEA building set. Standardization is key in augmenting the nuclear energy learning curve. The French did this in the 1970s to protect themselves from future OPEC embargos, thus securing both their economy and their environment, while raising both educational and living standards. Which brings us to the next of Shellenberger's seven points.

Four: Centralize construction with a single experienced builder

Building a nuclear reactor seems like a daunting task, and when you consider what has happened to several construction projects, it certainly looks daunting. However, there are companies and countries who can build their reactors on time and rather quickly. Think about construction times of as little as four years.

If one manages to standardize a design and create the off-the-shelf infrastructure required, one can build these reactors at record speeds. Additionally, the essential success factor is not so much Standardization or off-the-shelf but high-quality planning, supply chain management, and experience. Everything must be streamlined. The entire construction process depends on all the goods being delivered precisely on-time and construction workers must be able to go forward deliberately, finishing ahead of deadlines.

This point to me is most critical of all.

Five: Build as big as possible

This one should be obvious. To generate the most energy for as little money as possible, is to maximize profit potential, and therefore the incentive to build one in the first place. But I do not accept this one as gospel. There are plenty of practical situations imaginable that warrant building a smaller unit, or smaller units (SMRs). By chance, at COP23 in Bonn, Germany, I spoke to the Minister of Forestry, Environment, Climate Change and Natural Resources of The Gambia, Lamin Dibba, and when I asked him how many Gigawatts of capacity his country had, he told me that it was a couple of hundred Megawatt, and that it was based on burning oil. So, Gambia can decarbonize a part of their energy infrastructure. At this moment, they are looking for solar energy to bail them out. However, a nuclear reactor could do the same. But that would require a very cheap unit that would be competitive with solar, and it is questionable whether it can be done on such small scale. It doesn't make much sense to sell them a Gigawatt nuclear reactor unit. They would probably be better off building a couple small modular reactors rated at 150~300 MW a piece. This still fits the "build as big as possible" bill. But here you see that *big* is relative.

Also, we should build as big as reasonable. If we build 800-1000MW units, we have plenty of experience with this. If we build bigger, even the same design becomes a first of its kind (FOAK) e.g. the biggest-ever reactor vessel forging! The biggest-ever this or that... We have to be able to repeat and repeat often without pushing the envelope.

Six: Fix the Price, and don't allow changes during construction

We've now seen plenty of nuclear power projects with delays and revisions, which have made those projects much more expensive than initially thought. Think about the handful EPRs and AP1000s that are being built.

Once you have standardized your designs and have experience in licensing & building these reactors. you can sit down with your contractor and take through the usual time-table and discuss the price for the entire project. There may be discrepancies because of the layout of the terrain, but all in all, you should know exactly how much of everything you need to get the unit built.

And therefore, it should no longer be necessary to endlessly negotiate deals. A reactor unit build in Georgia should be 100% the same as a reactor unit built in New York. Down to the number of nuts and bolts.

Shellenberger notes: "The key to low-cost construction is low risk — not the estimated total cost. It is better for countries to go with a slightly more expensive builder who has significantly more experience — and who agrees to a fixed price in exchange for a no-changes rule — than one who offers a lower price at a "cost-plus" basis.

That's the key. The incentive for building the reactor within a set time, and for a fixed price should be with the builder. They can build in their own profit margins. And the customer is ensured that they will have a finished product on-time and at the agreed cost.

You see these kinds of arrangements everywhere in the construction business. Those who offer you estimated costs and do not commit to a stringent timetable quite often end up costing more money than those who commit to a fixed price and a set delivery date.

Seven: Finance with low-cost loans.

Given the fact that a nuclear power project doesn't generate money for perhaps several years (including pre-building time) means that you need loans to pay for the costs during non-operational project time. Note that wind/solar facilities not only receive subsidies, but their low capacity factors force added financing and construction of reliable backup systems to generate the power wind/solar don't and the wind/solar operators don't pay for that either.

I'll quote Shellenberger for this last one, as he said it best in his essay:

"Some of the highest costs that result from construction delays are simply paying the interest on loans. Avoiding high costs requires both avoiding delays and low-interest financing, whether from the government, the ratepayers (in the form of a fee on electric bills), or an international development bank. The riskiest phase of the project is in the planning, with

risk decreasing once construction has begun. Buyer countries should, therefore, have different financing for different phases. "

Shellenberger ends by saying that going with proven designs is safer.

I slightly disagree. I think that the inclusion of Molten Salt Reactors and Pebble Bed Reactors is warranted. Precisely because the amount of unearthed energy lying about in spent nuclear fuel pools and casks and depleted uranium repositories is a cash-opportunity of immense proportions. thorium waste from rare-earth mining is an even larger energy bank.

In conclusion, I believe that Michael Shellenberger has laid out a good strategy on how to make nuclear cheaper and quicker to deploy, without going into too much detail. I don't agree with everything he writes, I think there's still some margin of error left and right. But, as far as I am concerned, his arguments make sense. I have my own set of arguments to add, which we will consider in the next chapter.

Nuclear is ready for innovation

I don't want to give contemporary nuclear reactors bad press, because they don't deserve it, but we can do much better. And that's why I am so optimistic about nuclear energy in general. Those who claim that nuclear is an outdated technology are wrong. We're only 50 years behind schedule because we chose to go with light water reactors in the 1960s. Not only in terms of practical usefulness, but more so because we've only seen the tip of the iceberg. The solid fueled, water cooled nuclear reactor is only one of a multitude of designs possible. We have yet to unlock and supercharge the true potential of nuclear energy. First, we summarize some of the breakthroughs that are waiting for us, and then we're going to look at some of the real contenders in this race for new nuclear technologies—we must, China, Russia and others are.

Go passive! Nuclear Energy is intended to produce heat, which a working fluid transfers to a useful destination—turbine, processing plant, a community. After a reactor has been shut down, there's a grace period—a period of hours in which the reactor still generates a lot of heat because of the thermal mass of all its components, plus the radioactive decay of fission products. During this time, it needs to be cooled actively. We use pumps to circulate water through the reactor to cool it. To be fail-safe, we add redundancy in these safety systems by adding multiple pumps, backup generators, etc. This is an element that adds to the cost. But there are other, simpler ways of cooling a reactor based on the "always-on" principles of physics. For instance:

- Designing the core to be cooled by natural convection and water evaporation during the grace period, and convective air circulation to cool the reactor after the grace period has passed.

93

- We can use buffer coolants like lead or molten salts to extract heat from a reactor during the grace period.
- Use different coolant and fuel media. Preferably combine the two, as in a molten salt. In a Molten Salt Reactor, the fissile or fertile elements are dissolved in the salt. This greatly improves the reactor's thermal properties. Achieving higher temperatures and thus efficiencies at low pressures. Once the reactor gets shut down (scrammed) the heat capacity of the molten salt ensures that the reactor will stabilize itself. And, the molten salt can be drained into special tanks that ensure fission processes stop and the molten salt can even solidify (the opposite of a meltdown). Thus, making a reactor incident benign and easy to manage ORNL's MSRE salt is still in the same underground tanks to which it drained by gravity at shutdown in December 1969.

Standardize as much as possible! Do note that we can build our contemporary reactors within four years. So, that's not bad. However, on occasion we experience delays because many of the components used during construction are one-offs. And supply-chain management is very complex and can lead to delays if certain components do not meet changed specifications. This, we must avoid. We want standardized off-the-shelf components. We want the same electric generator for each unit. Preferably ones of which hundreds have been built already, with infrastructure able to build them as quickly as possible. We want this for every component, from the reactor core itself down to the nuts and bolts.

Modularize! Assembly is quicker than construction. Minimize construction time by creating a blueprint for the plant into which you can install all your modules. There are 600-MWe Reactor designs that are smaller than a Boeing 747. We can build airplanes on an assembly line. These airplanes consist of modules that are built separately and come together on the assembly line. A reactor is a lot less complex than an airplane, and the margins of error are the same. There's zero tolerance for mistakes. For reactors, how to modularize depends on the application, the working fluid, and the pressure during operation. Once we go low-pressure, I'm pretty sure that we could produce a fully assembled and tested reactor module a day on an assembly line. We move the assembled units to a plant site and install them in a waiting, ready-made cradle or hull. Other components could be assembled on factory lines as well. The more components we create this way, the bigger the deployment scale and the faster deployments become. Imagine being able to fabricate 219 Gigawatts of nuclear capacity in a year. We'd be able to decarbonize our world-wide electricity needs within 30 years. And this is accounting for one type of reactor design. As far as I can tell, there's room for multiple competing designs at high annual deployment rates.

Integrate! We can streamline installation time by including as many components within the module holding the reactor vessel as is optimal. For instance, think about integrating the heat exchangers and the pumps. Basically, create a closed module with a handful of input and output ports for measurements & controls, an exhaust port for extracting gaseous fission products, plus working-fluid inlets & outlets for the energy exchange between the reactor and the generator, or safety systems, or any process that will convert reactor heat into something useful. Note that this shouldn't be considered gospel because we also need to be able to replace or fix components as necessary. Over-integrating might make maintenance more difficult, even impossible. There are upsides and downsides with integrating all the components into one reactor module—integration must be done well and be tailored to particular reactor applications.

Liquify! First, by liquifying the fuel, we can eliminate the requirement of solid-fuel creation and management. This greatly reduces the complexity of the nuclear fuel cycle. We no longer have to create solid fuel pellets, thus streamlining the path from mine to reactor. Secondly, our so-called waste problem is a product of our choice to go with solid fuel. This means that after a period of several years fission products have accumulated inside the pellets and have rendered them inefficient for further use in the reactor. Also, the fuel pellets have been warped & cracked by buildup of gaseous fission products. The fuel pellets are considered spent. Some fission products absorb neutrons, inhibiting the chain reaction—the reactor could even approach shut down if spent fuel remained in the reactor long enough. A completely different fuel paradigm is to dissolve your fissile uranium and plutonium, or fertile thorium, and uranium into a liquid. Now your fuel is no longer encased by other fuel in a solid form, but it is bonded ionically and can move freely through the liquid. The great benefit here is that you have the optimum geometry that allows for 100% utilization of the fuel. Also, fission products like Xenon135 can now be extracted during operation. Ensuring that no neutrons get absorbed in non-fission reactions and thus being able to keep the reactor running for a much longer time, at higher fuel efficiencies. Switching from a solid fuel cycle to a liquid fuel cycle offers great benefits. This was successfully demonstrated with the first molten salt reactor at ORNL in the 1960s (see MSRE ref.).

Close the fuel cycle! In a nutshell? The only things that come out of a new-style fast-neutron reactor, are fission products. On a wider scale it would mean that the only things that come out of the civilian nuclear-power infrastructure are fission products. Consider the Russians for instance. They are the leading nation when fast breeder reactors are concerned. They have decades of experience and are the first to commercialize the BN800, which is a Liquid Metal Fast Breeder Reactor.

Closing the fuel cycle means that you will create and consume as much or more fuel than you put in, and that you make sure that all the fuel gets used. This principle was demonstrated at the Shippingport nuclear plant, in Pennsylvania, between 1977 and 1962, when the reactor core was converted to run on U233, bred from thorium.

Therefore, the only outputs you get from a reactor are fission products, some waste heat, and the energy we use for our day-to-day activities. Do note that these fission products are not all waste. They are only wasted if we chose to do so. And therefore...

Extract and separate fission products! A plethora of interesting and valuable radioactive and non-radioactive fission products are produced during nuclear fission. Think about cesium or technetium or molybdenum or zirconium or neodymium, just to name a few. Yet, we don't extract these. In the once-through fuel cycles many countries want to enact, these elements go to waste. After the fuel has been spent, they intend to bury it somewhere in a cave or in the desert. Naturally, I'm opposed to this practice. It's unwise to waste stuff that is more valuable than gold. Spent nuclear fuel is more valuable than gold because it contains energy, materials, medical isotopes, and isotopes for space exploration. We must move away from the solid and once-through fuel cycles. We must close our fuel cycles and extract all the valuables that are in there. We have proof of concept for some of the technologies that are required to achieve this. But we also have to perform a lot of research and development to close the knowledge gap and actually prove that we can extract and separate these valuable fission products. In any case, this is an opportunity, too big to waste.

Go ubiquitous! This last one is often overlooked. We live on a planet of limits. One of these limits is the concentration of certain elements available to us. My confidence in feasibility drops significantly if your technology requires rare elements for operation. Many wind generators for instance, require neodymium permanent magnets. And if they do not use permanent magnets, they use a bigger volume of precious copper. Some solar technologies require Indium, which is in pretty short supply. But there's also the high copper burden. For nuclear energy on the other hand, there are only two that really matter at this moment. One is niobium, which is a constituent of the zirconium Alloy used to create the fuel assemblies. And the other is Beryllium, which is a suggested, but not required, constituent of some of the salts proposed in Molten Salt Reactors. Beryllium is produced in exceedingly small quantities and might become an insurmountable hurdle when we want to scale up.

This might sound pessimistic, but I believe that the barrier for Beryllium is somewhere around 70 Gigawatts of added reactor capacity a year (at three tons of Be per GWe. Source: Lars Jorgensen—ThorCon Power). Another interpretation of going ubiquitous can be to use any fuel available and not be constricted by the relatively small volume of U235. We can use Th232, U235, and U238 as base fuels for nuclear power production. Additionally, we can use other fissile actinides that have been created in nuclear reactors. The volume of available fuel is great as soon as you step out of the U235 bubble. One of the benefits of including Th232 as a fuel is that it is a byproduct of the production of valuable rare earth metals and is considered a legal liability. By using Th232 commercially you would open more possibilities for heavy rare-earth extraction.

These are my eight secrets to cheap and abundant nuclear energy. I think they are complementary to Michael Shellenberger's seven secrets to cheap and abundant nuclear energy. Do note that we need Light Water Reactors to bridge the gap. We cannot afford to stop building these facilities just yet. In fact, it would be prudent to keep building these reactors up until the early 2030's. And when you consult the World Nuclear Association webpage you will see that this is going to happen. Regardless of the prohibitive sentiments against new nuclear development and innovation in the US.

Let's consider some of the startups that are designing new reactors according to some of the aforementioned principles.

NuScale

Imagine that you have a nuclear reactor that you can swap out, like a cartridge. NuScale is one of the first Light Water Reactor designs that comes with that promise and is almost commercially available. NuScale reactors are called NuScale Power Modules, and they are rated at 50 Megawatts electric, and a plant can hold 12 modules giving us a total rating of 600 Megawatts electric[107].

When we consider my summary of breakthroughs this one doesn't tick all the boxes. The NuScale concept does include Passive safety features, Modularization, integrated features and a high degree of standardization. It is more similar to our current generation of solid fuel reactors than some of the other reactors I will describe. It is also moving toward regulatory approval in the U.S., while other reactor builders are looking for approval abroad.

On January 9th, 2018, NuScale communicated that the US Nuclear Regulatory Commission (NRC) approved the NuScale design[108]. One element of the approval was the acknowledgement that a NuScale Power Module doesn't require any external safety-grade power source. This is a big deal. This is what made Fukushima happen—collapse of safety-grade power infrastructure, due to regulatory corruption. This was a unique circumstance, which forced Japan to replace its regulator[130]. We learned from it and adapted—the core principle of our very existence. Let's return to NuScale,

98

because it requires neither an external nor an internal safety-grade power infrastructure.

A NuScale Power Module is submerged in a pool of water. This pool is there to cool the reactor during the shutdown grace period. The water evaporates while cooling the reactor. After the water has evaporated, the water inside the reactor unit will evaporate and condense to maintain a continuous cycle of passive cooling, thus ensuring that the reactor is safe, without the requirement of external power for pumps and condensers, or intervention from reactor operators. You can find more information on the NuScale Website[109].

NuScale reactor units come equipped with the heat-exchangers and steam-generators integrated into the reactor unit. No steam leaves the module. The reactor modules are about 22 meters long and less than 3 meters wide, which means that they can be transported by trains, trucks and boats. Which offers a high degree of flexibility.

The small size of each unit, however, comes at a price. The LCOE for these units are comparatively high when contrasted to concepts such as the Molten Salt Reactor. That being said, they're still competitive and there will be a market for plants with one or more of these reactor modules.

Terrestrial Energy

If there's any chance of a first Molten Salt Reactor (MSR) becoming commercialized in the Western Hemisphere, Canada is probably your best bet. The regulatory pathways to commercialization in Canada are based on individual safety cases, rather than a standardized set of prerequisites that must be met as is the case in the US. In fact, the regulatory framework in the US for commercializing new designs is prohibitive if they are not based on the principles of the Light Water Reactor. And since the MSR principle is fundamentally different, it is unlikely that the US will be the country to have the first MSR going to market. And that's rather sad, because the Molten Salt Reactor is the brainchild of Alvin Weinberg (co-inventor of the LWR), who developed a practical MSR experiment[110] at Oak Ridge National Laboratory in Tennessee.

Despite repeated pleas from reactor designers on Capitol Hill, the US still hasn't managed to enact streamlined regulatory pathways for non-LWR designs. This is one of the reasons why I think that Terrestrial Energy might become the first to build a test reactor and commercialize it. The Canadian nuclear regulatory environment is much more accepting of new design principles. Terrestrial Energy might even come up ahead of China. But that's rather optimistic given the fact that China is aiming to have two test reactors built by 2020[111].

"The Chinese project has been funded by the central government and the two reactors are to be built at Wuwei in Gansu province, according to a statement on the website of the Chinese Academy of Sciences. The lead scientist on the project is Jiang Mianheng – the son of the former Chinese president Jiang Zemin – and it is hoped the reactors will be up and running by 2020."

This doesn't mean that these Chinese reactors will be ready to be commercialized. But it does show us that the Chinese are forging ahead with deliberate speed, and with access to all MSR research develo0pned at ORNL in the past.

Turning back to Terrestrial Energy. Terrestrial Energy's reactor design is called the IMSR, which is a shorthand for Integral Molten Salt Reactor. The name is reasonably self-explanatory. The IMSR is a reactor module that contains a graphite core, heat-exchangers, a control rod, and pumps. It ticks many good boxes: passive safety, integral components, modularization, a standardized design and liquid fuel.

The IMSR is designed as a thermal spectrum uranium burner, which means that neutrons are slowed down by a graphite moderator to split the U235 content of standard assay low-enriched uranium fuel and other fissile isotopes that emerge after neutron capture. The great benefit of these designs, is that they can be manufactured on assembly lines. They don't have to cope with high pressures, as in a Pressurized Water Reactor. In fact, an MSR runs at near atmospheric pressure. Once assembled, these reactors can be loaded on to trucks and moved to the site where they will be installed in a reactor silo, which is a subterranean concrete cylinder where the IMSR unit will be hooked up to a secondary heat exchanger (Remember, the primary heat exchanger is integral to the reactor unit). Most importantly, all the radioactive elements are sealed within the IMSR and never leave the reactor unit. This allows the construction of a very small and efficient containment structure, and greatly reduces the entire power plant's footprint, including for any waste storage.

Each IMSR unit comes with an operational lifespan of seven years. After which the unit will be left in the reactor silo to cool down. Meanwhile, a second IMSR unit has been loaded into an adjacent silo (there's no limit on how many operating silos can be built into a plant) and has been put into operation. Thus, ensuring seamless availability. During operation no fuel processing is required. New fuel salt is simply added periodically, no fuel is taken out, and gaseous neutron poisons like Xenon bubble out. Once the seven-year operational lifespan of the IMSR unit has expired, the remaining fuel salt is pumped into storage tanks, and the IMSR unit will be stored until it can be recycled. It is worth noting that neutron poisoning is not the limiting factor for operational lifespan; it is in fact the graphite moderator's condition.

The used fuel is still viable as reactor salt, perhaps with processing to remove any elements, like Hafnium, that could hurt the neutron economy. It could, in theory, be loaded into a new IMSR core, for another seven-year campaign, where it could continue to full burn-up, leaving only fission products behind. The reason is that the stability of fluoride salts chemically traps fission products and prevents them from creating chemical problems like plumbing corrosion. This is another exciting possibility for this technology: a reactor system that can consume 100% of its fuel, and 100% of its long-lived waste.

The International Atomic Energy Agency (IAEA) has created a 2016 status report on the IMSR that can be found on Terrestrial's website[112]

Terrestrial Energy has an excellent value proposition which you can see on their *"How it works"* page on their website—www.terrestrialenergy.com. The most interesting aspects are summarized in a schematic of the reactor and its appendages and benefits. The IMSR is not exclusively designed for electricity production, It can create heat to desalinate water, create synthesized transport fuels, provide process heat for industry, and even be used for thermal storage. Rauli Partanen, an independent energy author and energy analyst from Finland has written an excellent scenario[113] on how to decarbonize the entire infrastructure of a large city. It is called: "Decarbonizing Cities: Helsinki Metropolitan Area." In this scenario, ten 400MWth IMSR units are used to provide 8 TWh of heat, 12 TWh of electricity and 4 TWh of Hydrogen for a city of 1.5 Million people. Thus, decarbonizing every public service, including transport and centralized heating for all homes and buildings. If all of us would live in Helsinki style cities, it would take approximately 66,000 of these units to provide all the energy we need. That's about 12.5 Terawatts of electrical power capacity, and that's not really that much when we consider previous calculations. Note that in this case we would have thermal energy to boot, beyond anything that we've modelled in 100% renewable scenarios, so far.

This made me realize just how versatile a high-temperature reactor like the IMSR can be. For instance, in South Africa there's a water crisis of immense proportions. A prolonged and severe period of drought has all but eliminated the fresh water resources on which Cape Town depends[114,115]. Using Multistage flash evaporation desalination at 90 KWh[116] (thermal) per ~3800 liters an IMSR could potentially desalinate 400 Megaliters of water per day. Cape Town usually gets 24,000 Megaliters[117] a month, or 775 Megaliters per day. So, two IMSR400 units could theoretically be sufficient to desalinate all that water. Naturally, a careful balance must be struck between the usage of fresh and salt water resources and this means that there will be capacity left to generate electricity.

All of this comes at the low cost of up to 50 US dollars per Megawatthour[118]. And this is highly competitive when compared to modern Natural Gas fired power plants and onshore wind. I think that it is fair to conclude that Terrestrial Energy, with its IMSR400 design, is poised to grasp a sizeable chunk of tomorrow's market share in energy. We must remain patient for the time being as they are working towards this goal hand in hand with the Canadian Government.

The Molten Chloride Fast Reactor

I wanted to end with one development that might end up ticking almost all the boxes. The Molten Chloride Fast Breeder Reactor (MCFR)[119,120,121,122]. One of the best-known chlorides is table salt (sodium chloride). It turns out that if you heat this salt up, it becomes a liquid, but not only that. You can dissolve fissile and fertile elements such as thorium, uranium and plutonium into a molten sodium chloride medium. This salt doesn't moderate neutrons, which is a benefit if you're interested in closing the uranium-plutonium fuel cycle and doing it **in** the reactor without the need for a salt-processing plant.

Where the Molten Salt Reactor designed by Terrestrial is excellent, this design principle might become the ultimate in nuclear engineering. However, do note that Terrestrial is probably going to be first in delivering an actual product that will end up taking a very large share of the energy market. Let's consider the MCFR from the perspective of the mentioned breakthroughs, as designed by Elysium Industries and TerraPower (as a secondary option) the MCFR incorporates: passive safety, standardization, modularity, liquid fuels, a closed fuel cycle, fission-product extraction, and finally the use of ubiquitous materials and fuels. But all this comes at the cost of having much higher fissile content in core at all times, including startup, and no practical predecessor thus being entirely conceptual at this moment. So perhaps more testing is involved. The design by Terrestrial is based on the tried and tested moderated (thermal-neutron) Molten Salt Reactor Experiment which ran for thousands of hours back in the sixties. This offers a potential shortcut to the commercialization of the very first Molten Salt Reactor. And, following the MSRE's 1970s plans to implement a Th-U233 design (LFTR) leads to far less heavy actinide production (e.g., plutonium) and waste.

So far, not much can be said aside from the MCFR being a great concept—it consumes everything you throw into it, except most fission products. It is the ultimate dream for someone like me, who wants to see us utilize waste from prior nuclear energy production. I'm excited about the idea.

Another possible advantage of the MCFR is resistance to proliferation. Proliferation is tied to enrichment and fuel reprocessing. We can nail enrichment down by imposing stringent rules and continuous oversight and creating reactor designs that don't require highly enriched uranium. As for reprocessing, it is best if we can do everything inside the reactor itself. This is why fuel breeding inside the reactors (MCFR, LFTR) is preferable. In addition, a thermal-neutron reactor needs far less fissile (U233/235, Pu239) in core than does a fast reactor, including at startup. Of course, no reactor is needed to make a U235 bomb, to which various sovereign nations can testify.

In any case, there's enough reason to be optimistic. We've considered three contenders that are pushing for commercialization of new ways to extract energy from the nucleus. There are also other contenders out there like TerraPower (which is funded by Bill Gates), Holtec, U-Battery (no fission, just fission-product decay heat), ThorCon Power, X-Energy, GE-ARC-PRISM and Toshiba.

The boons of nuclear

It is time to consider all the advantages foregone thanks to the negative pressure that has been brought to bear on civilian nuclear energy. We keep ignoring these pressures when we're debating nuclear energy because the discussions are almost always focused on opponents misrepresenting it, and proponents offering rebuttals. To me, it feels like a contradiction. I've been studying energy matters for almost a decade now, and any negative effects from nuclear are grossly exaggerated, while benefits are intentionally muted. It is time to reverse this paradigm.

When a country considers building a nuclear power plant, it dedicates itself to a project that is going to last for the better part of a century. It's a huge responsibility to take on. Why is it going to last this long? How can the reactor be expected to last this long? Because of regulation and planning, the first phase of building the reactor requires due diligence by dozens of highly educated people. Next will be checks and balances performed by government regulators and possible financiers; Finally, if everything is approved, we get to the building phase which typically takes somewhere between 4 and 8 years, during which the plant itself does not generate any energy.

However, a nuclear plant under construction is a giant economic driver for the region where it is being built. During this phase the first real benefits of nuclear energy start to manifest themselves. Consider the following simple chain:

- We need steel, copper, lead, concrete, aluminium and other materials to build this nuclear power plant. People work to extract these, and because of this project they can keep working. This process involves the extraction and purification of these materials.

- Subsequently, the raw materials go to the manufacturing facilities, where they are turned into materials. Think about concrete, steel beams, wires & instrumentation, insulation & plumbing, and so on. Again, all this making sure people have long-lasting jobs and can even study for more advanced positions with more pay.

- Once the materials are ready to be used during the construction of the plant, they must be transported to the site by truckers and other transportation professionals.

- Finally, people at the construction site put all these materials together to build the nuclear power plant.

- And, while all that goes on, plant operators are being hired and trained.

This is a very shallow, one directional, one dimensional depiction of what goes on during the construction phase of a nuclear power plant. Suffice to say that the benefits in terms of permanent job creation are even more far reaching than we can imagine. There are work-chains running in all directions. Consider the need for large excavators and cranes for instance; or consider the sheer volume of economic activity that is generated by the workforce on the site in the first place. These people live regular lives, and have families, and hobbies and whatnot. They can contribute thanks to the good wages they earn on the construction site. In short, the nuclear power plant that is under construction is already a dynamo for economic growth and general welfare. To claim that a nuclear power plant is a money pit, or a black hole, as some do, is false. The money goes to people, and these people can lead meaningful lives thanks to the power plant being there. We will return to this notion again, later.

Let's consider this article[123] published in January 2018 on the Forbes website, written by James Conca: *"Nuclear power provides a whole lot more than just energy."* Conca highlights a recent analysis[124] by the Nuclear Energy Institute (NEI) called *"Economic impacts of the Columbia Generating Station."*

Conca writes: "Another study on the economics of nuclear energy found unequivocally that nuclear power plants provide substantial economic benefits to the states and regions in which they reside.

In this case, the study focused on the Columbia Generating Station, operated by Energy Northwest, in eastern Washington State. Performed by the Nuclear Energy Institute, the study found that the power plant, with a nameplate capacity recently increased to 1,207 MW, generates over 9 billion kWhs of emission-free electricity every year, enough to power the City of Seattle.

All while generating more than $690 million in economic output - over $475 million for Washington State and over $215 million for the rest of the United States."

What comes out of a nuclear power plant? Clean energy for economic activities that are worth money, wages for about a thousand people per Gigawatt, plus local tax revenue and spinoffs for an entire economy of suppliers for the plant. When you consider that there's an average wage between 60,000 & 130,000 USD/year for people working in the nuclear industry[125] in the United States, and that there are over 100,000[126] of them, who reinject at least 6 billion USD/year into the economy. Therefore, we may conclude that it's an industry that costs a lot of money, but that that money flows back into the economy through a well-paid, well-taken-care-of workforce of industry workers. What is more, these people all create extra jobs in healthcare, science, education, and the local economy. It's like and unlike the heydays of mining towns where precious metals and later stuff like coal would be mined in large quantities. While these towns have been left in the [coal] dust, nuclear comes with a completely different promise. First, it is unlikely that we will run out of nuclear fuel anytime soon, as I've shown you in earlier chapters. Second, nuclear power plants can operate many decades[127], and we can expand them as needed. Also, we can refurbish them to grant them another lease on life. It is common for nuclear power plants to get extensions on their operating licenses. It's because they are well-kept, and the owners and employees strive for operational excellence. Finally, once a terrain has been approved for nuclear operations, it is easier to build additional or replacement reactors, when needed.

107

Let's consider the jobs-impact of the Columbia nuclear power plant in the state of Washington as analyzed by the NEI[128].

Job impacts from Columbia Operations

Category	Washington	Rest of US	Total
Utilities	991	11	1002
Other Services (Except Public Administration)	397	183	580
Professional, Scientific, Technical Services	359	97	546
Retail Trade	207	61	268
Health Care, Social Assistance	198	108	306
Administrative, Waste Management Services	113	119	232
Accommodation, Food Services	104	38	142
Finance, Insurance	67	113	180
Real Estate, Rental and Leasing	59	39	98
Construction	55	4	59
Arts, Entertainment, Recreation	48	49	90
Wholesale Trade	47	43	97
Manufacturing	46	114	160
Transportation, Warehousing	39	54	93
Education Services	32	49	81
Other Industries	74	19	93
Total	2837	1102	3939

The boons of nuclear aren't exclusively environmental and economic. We also consider the creation of medical isotopes for diagnosis and treatment of ailments such as cancer; the creation of isotopes for the decontamination of the mail and food; the creation of isotopes for space-exploration.

And then there are the benefits of Generation IV reactors which offer even more versatility. Consider for instance the efficient use of process heat for: desalination through the Multi Stage Flash process; the use of process heat for the creation of synthetic carbon-neutral fuels; the use of process heat for chemical synthesis; district heating; steel-making; etc.

We've yet to see nuclear deployed in its most efficient way. Creating electricity is one of many things that civilian nuclear reactors do really well. And it would be wise to explore ancillary benefits of these facilities in greater detail, while simultaneously finding ways to use them for more purposes.

Let's not forget that we have roughly 440 nuclear reactors operational already and that these installations are being maintained with the highest degree of excellence. Keeping these installations up and running is paramount. This is zero-carbon electricity generation. Unfortunately, some power companies are shutting nuclear power plants down prematurely, which is often being heralded as good news by some environmentalists. But the opposite is true: closing nuclear energy capacity comes at high environmental costs.

Breaking the narrative

I've shown you that there are some deep-rooted falsehoods that must be overcome before this debate will end. It is in no-one's interest for this blind anti-nuclear narrative to linger on. Given the fact that I am a nuclear activist, I get drawn into these discussions all the time. And I must admit that it becomes tiring to have to go over the same issues time and time again. It feels like I've developed a narrative of my own. Which I think, is not good. Often, I feel compelled to start crusading against renewables, where there is no need. The truth is this: We don't have the luxury of lazily accepting feel-good promotions. If we want all people on this planet to live lives of plenty, in health and safety, while effectively reducing reliance on fossil fuels, we cannot blindly discount or accept any technology. It takes work to be serious about ending carbon emissions.

Let's keep our eyes on the ball. What's important? 100% renewables or zero emissions? Maximizing renewable deployment at harmful costs? Achieving our low emission goals and growing collectively as a civilization?

Researching our effects on the climate and the oceans is necessary. We've entered a critical phase in our planet's existence. We're now facing true calamities. Now is the time to make decisions that are based on sound judgement and science. This also means performing due diligence. Especially when someone makes extraordinary claims. Claiming that we can provide enough energy for mankind by deploying renewables (Solar and Wind) almost exclusively, is extraordinary. When such claims are exclusive in nature, it is especially important to scrutinize them, and all evidence presented in support such claims. We've noticed that there are almost always essential omissions and unwarranted assumptions—wishful thinking disguised as science.

Our political leaders have not managed to shape things the way they should be. If we consider the promises of the Kyoto Protocol and the Paris Agreement and contrast that with the practical actions of countries, with substantial carbon emissions, the evidence tells us that, except for France and very few others, no real progress has been made in all these years.

I look for leadership elsewhere. First, I consider the scientists—people who have the capabilities and means of considering all evidence and deduce what is happening to our planet and what our influence on it is. We see some of these scientists stepping up to enunciate that more needs to be done to settle the score with carbon emissions. Remember that fossil fuels are relatively cheap because their damages have not been calculated into their price. Two things must happen in a relatively short time:

1. Enact a global *carbon fee and dividend* system. The price of carbon-based fuels must increase progressively, and the money that is collected should be redistributed among the people evenly. Economists and scientists have researched its possible effects and deemed it one of the most effective means of putting the use fossil fuels (and their emissions) into decline[129].

2. Nuclear power must be included in the suite of solutions to Climate Change. Therefore, nuclear should be an integral part of the discussion, represented by innovative startups, power companies, industry organizations, and research institutes. We best leave the word renewable out of the equation, because not a single source of energy can be harvested indefinitely and entirely renewable. Not even wind and solar. But sustainable is a better word to use here. We can extract energy from the nucleus in a sustainable way. We can apply circular principles to all energy technologies involved, trying to re-use as many precious materials as possible. But that will require more energy, we can recycle almost everything if only we have the energy to do it. If we can strike a careful balance between nuclear and hydro, geothermal, wind, and solar we may be able to stop the emissions from fossil fuels just in time to keep the damage to our biosphere manageable.

This has been James Hansen's staple suggestion in almost all his communications regarding how to address climate change effectively. When implemented correctly, all low-carbon energy sources will automatically become more competitive with coal, gas and oil. The incentive to take the cheaper energy source over the more expensive one is evident. This is one of the few suggestions that will catalyze sustainable development based on free market economics.

An additional incentive could come from governments by catalyzing innovation through collaboration between different national laboratories and private companies, catalyzed by innovation funds.

Finding people who have either transcended or always stood above these narratives is easy, they are everywhere. Consider for instance, Bill Gates, whose philanthropy and investments are aimed at the betterment of humanity. Not only is he, together with his wife Melinda, very involved in the deliberate and calculated distribution of vaccines amongst other things, he has also invested in new nuclear developments. TerraPower, for instance, is such a company that is working on reactors that can consume spent nuclear fuel.

I consider French Leader Emmanuel Macron as someone who has—recently and successfully—transcended the staple anti-nuclear environmentalist narrative. For he has come out in support of nuclear after becoming French President. First, he was completely inundated with the 100% renewable narrative. But after consideration of the facts, he has changed his tune considerably. Now he is clear, he won't make the same mistake as the Germans and doesn't rule out that France is going to build new reactors. This doesn't rule out that old reactors in France may be shut off however.

Michael Shellenberger, founder and director of Environmental Progress—one the most inspiring people I know—has transcended the narrative. Despite coming from a *hippy* background with a track record in environmental activism, he has found the moral and transcendent purpose of nuclear power by careful investigation of the facts. He is now (March 2018) running as a gubernatorial candidate in California and gaining momentum. Why do I hope he gets elected? First, he is an absolute environmentalist who understands the pivotal role of nuclear in a clean world. As a humanist, he cares deeply for others and wants to create a better world for all. He is looking to reinvigorate the housing sector to eliminate the scourge of homelessness that has taken hold of California; he also knows the intrinsic value of good education and high-tech jobs. And will fight to keep Diablo Canyon open. One of the best run nuclear power plants in the world.

When we look further we note expert scientists and thinkers like Steven Pinker, Brian Cox, Ken Caldeira, James Hansen, Stephen Tindale (died in 2017), Steven Chu, David Mackay (died in 2016), and Ernest Moniz (Who recently joined the advisory board of Terrestrial Energy).

The humanist case for nuclear energy

A friend of mine asked what my personal case for nuclear energy was. And I wondered if I could fit this in a nutshell as good as possible. So here comes the humanist case for nuclear energy.

The cost of energy poverty is death. Up to a certain degree this energy poverty is a problem for at least a billion to 2 billion people—another 2~3 billion people are coming due to unstoppable population growth. Too many of these people suffer and eventually die from malnutrition and a lack of clean water for hygiene and other purposes.

We need more low-carbon, clean, high-tech energy to lift billions of people out of different levels of poverty, while effectively addressing climate Change. I want all people on the planet to live in general prosperity, safety, and happiness. High energy and reliability per capita correlates with better living standards, and this is underlined by academic studies. Before the turn of this century, we will need more energy to provide basic needs like water, food, shelter, and healthcare for an additional 2 to 3 billion people.

The reason why I say high-tech, is because it will provide high-intelligence and high-education spinoffs. The general population of a country will become more knowledgeable and scientifically literate as educated people are in greater demand and everything is done to educate them. If implemented carefully, with a bottom-up strategy in mind, everybody could benefit from a nation's goal to reach new heights. Each child should be well-nourished. Not just in terms of food, water, and love. But also, in terms of education. The US Apollo program did exactly this!

Sciences like Physics, Chemistry, and Biology should be a staple in schools. And this should be reflected on what is available on TV and the internet as well. Remember those great documentaries by Carl Sagan and David Attenborough? I am confident that an increase in the overall availability of education and science will lead to a healthier and more optimistic society. But don't forget that we need caretakers, dancers, musicians, and artisans too.

Having read this book, it should have become clear why I am a fierce advocate for nuclear power. I hope you will join me as an advocate for clean and reliable power. Many people must be turned away from the dogmatic insistence that renewables will do the job we must do for our descendants' planet. We have a lot of fact advocacy to do. That is not to say that renewables are useless, but the hubris surrounding these technologies must be deflated. I am convinced that the brunt of our energy must come from nuclear, hydro and geothermal.

Disregarding concerns enunciated by environmental experts from all over the world while continuing some quest to extirpate civilian nuclear energy from any future scenario is reckless in the extreme. So far, wind and solar have managed to crest the first pebble at the foot of the emissions mountain despite decades of massive investments.

Remember, with science and reason as your guides you will become intellectually satisfied. However, this also means that people will disagree with you strongly. As it stands, we're a niche-bunch with a mission. We must teach people to think for themselves again and step outside the relative safety of the group and its consensus. Once we achieve critical mass, we can turn this situation around and start building a fairer society with an optimistic future for all.

Appendix

References

My references are [almost] exclusively online because I want you to be able to have access to all the sources I've used. I want to show you that you can find reliable information online. But... Always remain skeptical and accept evidence only from reputable sources [after careful examination, if possible]. It takes some effort to weed out the nonsense, one of the problems when stepping outside the realm of peer-review.

This does not equal an academic bibliography. However, it is an attempt to show you how to source evidence to support your claims, or where to find interesting quotes

1   *Projections of Climate Change, Climate Sensitivity, Cumulative Carbon—Reto Knutti 2013*
    www.ipcc.ch/pdf/unfccc/cop19/2_knutti13sbsta.pdf

2   *NASA: Climate Change: How do we Know*
    climate.nasa.gov/evidence/.

3   *Greater future global warming inferred from Earth's recent energy budget—Patrick T. Brown & Ken Caldiera 2017*
    www.nature.com/articles/nature24672

4   *NOAA: Is sea level rising?*
    oceanservice.noaa.gov/facts/sealevel.html

5   *NOAA: Sea level rise viewer*
    coast.noaa.gov/slr/

6   *Holocene Extinction*
    en.wikipedia.org/wiki/Holocene_extinction

7   *Worldatlas: Timeline of mass extinction events on Earth*
    www.worldatlas.com/articles/the-timeline-of-the-mass-extinction-events-on-earth.html

8   *Declining oxygen in the global ocean and coastal waters—Denise Breitburg et al 2018*
    science.sciencemag.org/content/359/6371/eaam7240

9   *NOAA: What is a dead zone?*
    oceanservice.noaa.gov/facts/deadzone.html

10 *Global carbon budget 2017—Le Quéré et al 2017*
    www.globalcarbonproject.org/carbonbudget/17/files/GCP_CarbonBudget_2017.pdf

11 *OECD.Stat: Greenhouse gas emissions*
    stats.oecd.org/Index.aspx?DataSetCode=AIR_GHG

12 *EIA: International Energy Outlook 2017*
    www.eia.gov/outlooks/ieo/pdf/04842017.pdf

13 *IPCC: Climate Change 2001: Synthesis Report*
    www.ipcc.ch/ipccreports/tar/vol4/011.htm

14 *United Nations Climate Change: Kyoto Protocol*
    unfccc.int/kyoto_protocol/items/2830.php

15 *Embassy of France in Washington: Nuclear Energy in France*
    ambafrance-us.org/spip.php?article637

16 *NASA: NASA-MIT Study evaluates efficiency of oceans as heat sink, atmospheric gas sponge*
    climate.nasa.gov/news/2598/nasa-mit-study-evaluates-efficiency-of-oceans-as-heat-sink-
    atmospheric-gases-sponge/

17 *AAAS Science, Canfield & Kump, vol 339, p533, 2/1/2013*

18 *NOAA: Hawaii carbon dioxide time series*
    www.pmel.noaa.gov/co2/file/Hawaii+Carbon+Dioxide+Time-Series

19 *Long-term response of oceans to CO_2 removal from the atmosphere*
    www.nature.com/articles/nclimate2729

20 *Sciencemag: Almost all of the 29 coral reefs on UN Heritage list damaged by bleaching*
    www.sciencemag.org/news/2017/06/almost-all-29-coral-reefs-un-world-heritage-list-
    damaged-bleaching

21 *Smithsonian National Museum of Natural History: Zooxanthellae and Coral Bleaching*
    ocean.si.edu/slideshow/zooxanthellae-and-coral-bleaching

22 *Acidic Oceans: Why Should We Care? —Perspectives on Ocean Science*
    www.youtube.com/watch?v=kQMZfCKuFIQ

23 *A short history of ocean A short history of ocean acidification science in the 20th century: a chemist's view—Biogeosciences, 10, 7411–7422, 2013*
    www.biogeosciences.net/10/7411/2013/bg-10-7411-2013.pdf

24 *IPCC: Fifth Assessment Report (AR5)*
    www.ipcc.ch/report/ar5

25 *Target atmospheric CO_2: Where should humanity aim?*
    benthamopen.com/ABSTRACT/TOASCJ-2-217

26 *How does it feel? —James Hansen 2017*
    www.columbia.edu/~jeh1/mailings/2017/20171006_NorthDakota.pdf

28 *Carbon Dioxide and Nuclear Energy: The Great Divide and How to Cross It by Meredith Angwin*
    www.theenergycollective.com/meredith-angwin/92451/carbon-dioxide-and-nuclear-energy-
    great-divide-and-how-cross-it

29   *World total primary energy consumption by region, 2015-50*
www.eia.gov/outlooks/ieo/excel/ieotab_1.xlsx

30   *World Population by region, reference case, 2015-50*
www.eia.gov/outlooks/ieo/excel/appj_tables.xlsx

31   *Jacobson et al., Joule 1, 108-121, 2017*
web.stanford.edu/group/efmh/jacobson/Articles/I/CountriesWWS.pdf

32   *EIA: Electric Power Monthly*
www.eia.gov/electricity/monthly/epm_table_grapher.php?t=epmt_6_07_b

33   *REN21, Renewables 2017 Global Status Report*
www.ren21.net/wp-content/uploads/2017/06/17-8399_GSR_2017_Full_Report_0621_Opt.pdf

34   *The Shift Project Data Portal: Historical Electricity Installed Capacity Statistics*
www.tsp-data-portal.org/Historical-Electricity-Capacity-Statistics#tspQvChart

35   *Vestas: Material Use Turbines, version 1.0, December 2014*
www.vestas.com/~/media/vestas/about/sustainability/pdfs/material%20use%20brochure%20v
1%20dec%202014.pdf

36   *SunPower: X-series Residential Solar Panels X22-360*
us.sunpower.com/sites/sunpower/files/media-library/data-sheets/ds-x22-series-360-residential-
solar-panels.pdf

37   *US Geological Survey, Mineral Commodity Summary, January 2018, Copper*
minerals.usgs.gov/minerals/pubs/commodity/copper/mcs-2018-coppe.pdf

38   *Rare Earth Elements: Overview of Mining, Mineralogy, Uses, Sustainability and
environmental impact—Nawshad Haque, Anthony Hughes, Seng Lim, Chris Vernon 2014*
www.mdpi.com/2079-9276/3/4/614/pdf

39   *Vestas, (2012). Life Cycle Assessment of Electricity Production from an onshore V90-3.0 MW
Wind Plant – 30 October 2013, Version 1.1*
www.vestas.com/~/media/vestas/about/sustainability/pdfs/lca_v903mw_version_1_1.pdf

40   *Zhang, Y., 2013: Peak Neodymium – Material Constraints for Future Wind Power*
www.diva-portal.org/smash/get/diva2:668091/FULLTEXT01.pdf

41   *Life Cycle Assessment of 1 KWh energy generated by Gamesa G114-2.0MW On-shore wind
farm—Sergio Rodriguez Carrascal 2014*
www.siemensgamesa.com/recursos/doc/productos-servicios/aerogeneradores/life-cycle-
assesment-g114-20-mw.pdf

42   *Life Cycle Assessment of an innovative recycling process for crystalline silicon photovoltaic
panels.*
www.sciencedirect.com/science/article/pii/S0927024816001227

43   *Life cycle assessment of utility-scale CDTE PV Balance of Systems—Parakhit Sinha &
Mariska de Wild-Scholten 2012*
smartgreenscans.nl/publications/Sinha-and-deWildScholten-2012-Life-cycle-assessment-of-
utility-scale-CdTe-PV-Balance-of-Systems.pdf

44   *Roadmap to Nowhere—Mike Conley & Dr. Timothy Maloney 2017*
www.roadmaptonowhere.com/

45   *Burden of proof: A comprehensive review of the feasibility of 100% renewable-electricity*

systems—*Ben Heard et al 2016*
www.sciencedirect.com/science/article/pii/S1364032117304495

46   *Evaluation of a proposal for reliable low-cost-grid power with 100% wind, water, and solar— Christopher Clack et al 2016*
www.pnas.org/content/114/26/6722.full.pdf

47   *Low-cost solution to the grid reliability problem with 100% penetration of intermittent wind, water, and solar for all purposes—Jacobson et al 2015*
www.pnas.org/content/112/49/15060#T1

48   *Supporting Information—Low-cost solution to the grid reliability problem with 100% penetration of intermittent wind, water, and solar for all purposes—Jacobson et al 2015*
www.pnas.org/content/pnas/suppl/2015/11/19/1510028112.DCSupplemental/pnas.1510028112.sapp.pdf

49   *MZJ Hydro Explainer—Anonymous 2018*
kencaldeira.wordpress.com/2018/02/28/mzj-hydro-explainer/

50   *Xtools—Wikipedia Information page for Mark Z. Jacobson*
xtools.wmflabs.org/articleinfo/en.wikipedia.org/Mark_Z._Jacobso

51   *There is a new form of climate denialism to look out for – so don't celebrate yet—Naomi Oreskes 2015*
www.theguardian.com/commentisfree/2015/dec/16/new-form-climate-denialism-dont-celebrate-yet-cop-21

52   *Hansen, Emanuel, Wigley, Caldeira: Energy for Humanity Press Conference COP21*
www.youtube.com/watch?v=ZlbziE-78DI

53   *Department of Energy: Energy Dept. finds major potential to grow clean, sustainable US Hydropower*
energy.gov/articles/energy-dept-report-finds-major-potential-grow-clean-sustainable-us-hydropower

54   *Wikipedia: List of Nuclear Reactors*
en.wikipedia.org/wiki/List_of_nuclear_reactors

55   *NASA: Coal and gas are far more harmful than nuclear power—Pushker Kharecha & James Hansen 2013*
climate.nasa.gov/news/903/coal-and-gas-are-far-more-harmful-than-nuclear-power

56   *World Nuclear Association: Military Warheads as a source for Nuclear Fuel*
www.world-nuclear.org/information-library/nuclear-fuel-cycle/uranium-resources/military-warheads-as-a-source-of-nuclear-fuel.aspx

57   *Foods known to contain naturally occurring Formaldehyde*
www.cfs.gov.hk/english/whatsnew/whatsnew_fa/files/formaldehyde.pdf

58   *REN21, Renewables 2017 Global Status Report*
www.ren21.net/wp-content/uploads/2017/06/17-8399_GSR_2017_Full_Report_0621_Opt.pdf

59   *The New York Times: Germany's shift to Green Power Stalls, Despite Huge Investments— Stanley Reed 2017*
www.nytimes.com/2017/10/07/business/energy-environment/german-renewable-energy.html

60   *Dr. James Hansen COP23 Interview*

www.youtube.com/watch?v=a0MsAs-qCSY&t

61   *NASA: Far Northern Permafrost may unleash carbon within decades*
     www.nasa.gov/feature/jpl/far-northern-permafrost-may-unleash-carbon-within-decades

62   *UN Population Division: World Population Prospects 2017*
     esa.un.org/unpd/wpp/Maps/

63   *Our world in data: Share of the world population living in Absolute Poverty 1820-2015*
     ourworldindata.org/wp-content/uploads/2013/05/World-Poverty-Since-1820.png

64   *Lawrence Livermore National Laboratory: Energy Flow Charts*
     flowcharts.llnl.gov

65   *Lawrence Livermore National Laboratory: Estimated US Consumption in 2016*
     flowcharts.llnl.gov/content/assets/images/energy/us/Energy_US_2016.png

66   *Social Progress Index 2017*
     www.socialprogressindex.com

67   *Wikipedia: List of countries by energy consumption per capita*
     en.wikipedia.org/wiki/List_of_countries_by_energy_consumption_per_capita

68   *Advanced Combustion Engines—Coursework—Christopher Goldenstein 2011*
     large.stanford.edu/courses/2011/ph240/goldenstein2/

69   *EU Science Hub: Well-to-Wheels Analyses*
     ec.europa.eu/jrc/en/jec/activities/wtw

70   *Phys.org: Why a hydrogen economy doesn't make sense*
     phys.org/news/2006-12-hydrogen-economy-doesnt.html

71   *Renewable Power-To-Gas: A technological and economic review—Manuel Götz et al.,
     Renewable Energy Volume85, 1371-1390, January 2016*
     www.sciencedirect.com/science/article/pii/S0960148115301610

72   *World Health Organization: 7 Million premature deaths annually linked to air pollution*
     www.who.int/mediacentre/news/releases/2014/air-pollution/en/

73   *EIA International Energy Outlook 2017*
     www.eia.gov/outlooks/ieo/pdf/04842017.pdf

74   *EIA: Electric Power Monthly*
     www.eia.gov/electricity/monthly/epm_table_grapher.php?t=epmt_6_07_b

75   *Vestas Material Use Turbines, Version 1.0, December 2014*
     www.vestas.com/~/media/vestas/about/sustainability/pdfs/material%20use%20brochure%20v
     1%20dec%202014.pdf

76   *SunPower X-Series Residential Solar Panels X21-335-BLK*
     us.sunpower.com/sites/sunpower/files/media-library/data-sheets/ds-x21-series-335-345-
     residential-solar-panels.pdf

77   *Smartgrid.gov: What is the Smart Grid?*
     www.smartgrid.gov/the_smart_grid/smart_grid.html

78   *Providing all global energy with wind, water, and solar power, part II: Reliability, system and
     transmission costs, and policies—Mark Z Jacobson & Mark A Delucchi 2010*
     web.stanford.edu/group/efmh/jacobson/Articles/I/DJEnPolicyPt2.pdf

79   *EIA: Natural gas-fired combustion turbines are generally used to meet peak electricity load*

www.eia.gov/todayinenergy/detail.php?id=13191

80 *Department of Energy: How gas turbine power plants work*
    energy.gov/fe/how-gas-turbine-power-plants-work

81 *Robert Hargraves – Thorium Energy Cheaper than coal @ THEC12*
    www.youtube.com/watch?v=aylyiVua8cY

82 *Normal Cubic Meter of Natural Gas conversion chart*
    www.convert-me.com/en/convert/energy/cmsgas.html?u=cmsgas&v=1

83 *Erneuerbare Energien und klimaschutz.de: Specific Carbon Dioxide Emissions of various fuels*
    www.volker-quaschning.de/datserv/CO2-spez/index_e.php

84 *Agora Energiewende: Agorameter*
    www.agora-energiewende.de/en/topics/-agothem-/Produkt/produkt/76/Agorameter/

85 *Wikipedia: Neodymium Magnets*
    en.wikipedia.org/wiki/Neodymium_magnet

86 *Material Flows Resulting from Large Scale Deployment of Wind Energy*
    www.mdpi.com/2079-9276/2/3/303/pdf

87 *Wikipedia: Monazite*
    en.wikipedia.org/wiki/Monazite

88 *Institute for Energy Research: Big wind's dirty little secret: Toxic lakes and radioactive waste*
    instituteforenergyresearch.org/analysis/big-winds-dirty-little-secret-rare-earth-minerals

89 *The Conference Board: International Comparisons of Hourly compensation costs in manufacturing*
    www.conference-board.org/ilcprogram/index.cfm?id=28277

90 *United States Nuclear Regulatory Commission: Source Material*
    www.nrc.gov/materials/srcmaterial.html

91 *ZDnet: Solve the Energy and Rare Earth crisis: Join the Thorium Bank*
    www.zdnet.com/article/solve-the-energy-and-rare-earth-crisis-join-the-thorium-bank/

92 *OECD: Uranium 2016: Resources, Production and demand*
    www.oecd-nea.org/ndd/pubs/2016/7301-uranium-2016.pdf

93 *Uranium Extraction from seawater—coursework—Ken Ferguson 2012*
    large.stanford.edu/courses/2012/ph241/ferguson2/

94 *Collection of uranium from seawater—Masao Tamada 2009*
    www.iaea.org/OurWork/ST/NE/NEFW/documents/RawMaterials/TM_Vienna2009/presentations/22_Tamada_Japan.pdf

95 *World Nuclear Association: Thorium*
    www.world-nuclear.org/information-library/current-and-future-generation/thorium.aspx

96 *Scientific American: The working of an ancient Nuclear Reactor*
    www.scientificamerican.com/article/ancient-nuclear-reactor/

97 *Nature: Why Finland now leads the world in nuclear waste storage*
    www.nature.com/news/why-finland-now-leads-the-world-in-nuclear-waste-storage-1.18903

98 *Wikipedia: KBS-3 storage*
    en.wikipedia.org/wiki/KBS-3

99   *Why nuclear programs rarely lead to nuclear proliferation*
       www.mitpressjournals.org/doi/abs/10.1162/ISEC_a_00293

100   *World Nuclear Association: Uranium Enrichment*
       www.world-nuclear.org/information-library/nuclear-fuel-cycle/conversion-enrichment-and-
       fabrication/uranium-enrichment.aspx

101   *Wikipedia: Magnox*
       en.wikipedia.org/wiki/Magnox

102   *Wikipedia: RBMK*
       en.wikipedia.org/wiki/RBMK

103   *World Nuclear Association: Plutonium*
       www.world-nuclear.org/information-library/nuclear-fuel-cycle/fuel-recycling/plutonium.aspx

104   *Wikipedia: Megatons to Megawatts*
       en.wikipedia.org/wiki/Megatons_to_Megawatts_Program

105   *Wikipedia: BN800 Reactor*
       en.wikipedia.org/wiki/BN-800_reactor

106   *OECD: Projected Costs of Generating Electricity: 2010 Edition*
       www.oecd.org/about/publishing/45528378.pdf

107   *NuScale Power: How NuScale Technology Works*
       www.nuscalepower.com/our-technology/technology-overview

108   *NEI: NuScale Makes history with filing for SMR Design Approval*
       www.nei.org/news/2017/nuscale-makes-history-with-filing-for-smr-design

109   *NuScale Power: NuScale Technology Innovations*
       www.nuscalepower.com/our-technology/design-advances

110   *Wikipedia: Molten Salt Reactor Experiment*
       en.wikipedia.org/wiki/Molten-Salt_Reactor_Experiment

111   *NextBigFuture: China spending US$3.3 Billion on Molten Salt nuclear reactors for faster
       aircraft carriers and in flying drones*
       www.nextbigfuture.com/2017/12/china-spending-us3-3-billion-on-molten-salt-nuclear-
       reactors-for-faster-aircraft-carriers-and-in-flying-drones.html

112   Terrestrial Energy: Status Report—IMSR400
       www.terrestrialenergy.com/wp-content/uploads/2018/02/IMSR400.pdf

113   *Energy for humanity: Finnish cities to investigate the potential for small nuclear reactors to
       decarbonize district heating*
       energyforhumanity.org/en/news-events/news/news/finnish-cities-investigate-potential-small-
       nuclear-reactors-decarbonize-district-heating/

114   *BBC: Will Cape Town be the first city to run out of water?*
       www.bbc.com/news/business-42626790

115   *CNN: Cape Town rejoices as rain falls on drought-stricken city*
       edition.cnn.com/2018/02/11/africa/cape-town-rain-day-zero-intl/index.html

116   *Energy Cost of Desalination—Coursework—Jesse Sherer 2010*
       large.stanford.edu/courses/2010/ph240/sherer2/

117   *How many days of water does cape town have left*
       www.howmanydaysofwaterdoescapetownhaveleft.co.za/how-much-water-is-that.html

118 *Terrestrial Energy: IMSR Power plants are a cleaner and cost-competitive alternative to burning fossil fuels*
www.terrestrialenergy.com/technology/competitive/

119 *TerraPower: MCFR Solutions: Nuclear innovation for new options in American industry*
terrapower.com/technologies/mcfr

120 *TerraPower: On Molten Salt Reactors*
terrapower.com/updates/on-molten-salt-reactors/

121 *Oak Ridge National Laboratory—2015 Workshop on Molten Salt Reactor Technologies*
public.ornl.gov/conferences/MSR2015/Presentations.cfm

122 *TerraPower: Presentation—TerraPower and the Molten Chloride Fast Reactor*
public.ornl.gov/conferences/MSR2015/pdf/16-151015%20-%20MCFR%20at%20TerraPowerJeffLatkowski.pdf

123 *Nuclear power provides a whole lot more than just energy*
www.forbes.com/sites/jamesconca/2018/01/16/nuclear-power-provides-a-whole-lot-more-than-just-energy/#6dd79e605e7e

124 *Economic impacts of the Columbia Generating Station*
www.energy-northwest.com/ourenergyprojects/Columbia/Documents/NEI_EconomicImpacts-ColumbiaGeneratingStation-010918.pdf

125 *Payscale: Nuclear Engineer Salary*
www.payscale.com/research/US/Job=Nuclear_Engineer/Salary

126 *NEI: Factsheets*
www.nei.org/master-document-folder/backgrounders/fact-sheets/job-creation-and-economic-benefits-of-nuclear-ener?feed=factsheet

127 *International Atomic Energy Agency: Going long term: US Nuclear power plants could extend operating life to 80 years*
www.iaea.org/newscenter/news/going-long-term-us-nuclear-power-plants-could-extend-operating-life-to-80-years

128 *NEI: Economic Impacts of the Columbia Generating Station, January 2018*
www.energy-northwest.com/ourenergyprojects/Columbia/Documents/NEI_EconomicImpacts-ColumbiaGeneratingStation-010918.pdf

129 *Climate Science, Awareness and Solutions: Environment and Development challenges: The imperative of Carbon Fee and Dividend—James Hansen 2015*
csas.ei.columbia.edu/2015/11/11/environment-and-development-challenges-the-imperative-of-a-carbon-fee-and-dividend/

130 www.nirs.org/fukushima/naiic_report.pdf

About the author

Mathijs was born in Geleen, The Netherlands, and still lives there. He graduated from Hogeschool Zuyd in 2003 where he studied information technology. Before becoming a writer in 2012, Mathijs was independent in all his endeavors, he worked for multinational *Philips*, communication technology company *Royal KPN*, and pension fund investor *APG* (3rd largest in the world) as an IT specialist. Mathijs also owned a business for four years, again specializing in IT. In 2009 Mathijs started working on a new scouting method for a professional sports organization, which was based on defining metrics and performing statistical analyses, the culmination of which was a noticeable increase in attendance and overall scoring ability. In 2012, driven by a keen interest in science and technology, Mathijs decided to switch to writing about energy and climate change. His prime motivation for writing about these matters comes from a sense of responsibility: to leave this planet a better place than he found it.

In 2015 Mathijs published *Highway to Dystopia*—about climate change, energy, politics, economics, and religion; in 2016 he published *Science a la Carte*—about climate change and energy; And *the Non-Solutions Project*—a rebuttal to the 100% WWS Roadmap by Jacobson et al.

In 2017 Mathijs started a YouTube Channel called *"The Nuclear Humanist."* If you want to support him in his endeavors, please consider becoming a Patron at **Patreon.com/thenuclearhumanist**.

In 2017 Mathijs was a guest-speaker in the Reason and Science Coffeeshop in Watford (UK), and at the eighth Thorium Energy Alliance Conference in Saint Louis (US).

Made in the USA
San Bernardino, CA
10 June 2018